IS HEATHCLIFF A MURDERER?

IS HEATHCLIFF A MURDERER?

PUZZLES IN 19TH-CENTURY FICTION

JOHN SUTHERLAND

QUALITY PAPERBACK BOOK CLUB
NEW YORK

This edition was especially created in 2001 for Quality Paperback Book Club by arrangement with Oxford University Press, Inc.

Printed in the United States of America

Contents

Introduction and Acknowledgements

> 'Since the little puzzle amuses the ladies, it would be a pity to spoil their sport by giving them the key.'
>
> (Charlotte Brontë, about two female correspondents who had written to her publisher inquiring about the fate of Paul Emanuel in *Villette*)

Personally, I have always thought 'how many children had Lady Macbeth?' a perfectly good question. I am also curious about how old Hamlet is, what subjects he studied at the University of Wittenberg, and what grades he got from his teachers there. *Is Heathcliff a Murderer?* explores that forbidden territory, the *hors texte*—or, more precisely, that implied and ambiguous world which lies on the other side of the words on the page.

The substance of this book originates in the annotation of Victorian novels for various 'classic reprint' series, and, more specifically, that with which I have been most closely connected, OUP's revived paperback line of 'World's Classics'. There are tens of thousands of readers of this (and similar) series and the mass of such readers do not, I suspect, much worry about Deconstruction, New Historicism, or the distinction between extradiegesis and intradiegesis, But they do wonder, in their close-reading way, whether Becky killed Jos, exactly what nationality Melmotte is, what the 'missile' is that Arabella Donn pitches at Jude Fawley's head, what Heathcliff does in the three years which sees him leave Wuthering Heights a stable-boy and return a gentleman, and what Paul Emanuel does in *his* three years' sojourn in Guadaloupe.

What follow are literary brain-teasers. In the spirit of

those who enjoy such games, I have let my ingenuity rip in places. I would not go to the stake for some of the readings offered here, and I have no doubt that many readers of World's Classics will come up with cleverer and more plausible solutions than mine. But I would argue that however far my solutions are fetched the problems which inspire them are not frivolous. It *is* worthwhile for readers to be curious where Sir Thomas Bertram's wealth comes from, or to wonder why *The Picture of Dorian Gray* is so 'queerly' disturbing, or to inquire why George Eliot and Henry James consciously flawed the printed endings to their greatest novels. It is less crucial, but no less thought-provoking, that Henry Esmond—the highly literate creation of a highly literate author—should quote from a work forty years before it will be written. The questions which have provoked this book are, I maintain, good questions.

I have a number of debts to acknowledge. David Lodge (who adapted the novel for television) drew my attention to the striking seasonal anomaly in *Martin Chuzzlewit*. Rosemary Ashton corrected the *Middlemarch* chapter radically. Alison Winter helped with the 'science' in the *Jane Eyre* and *Oliver Twist* chapters. Philip Horne pointed out to me R. H. Hutton's meddling review of *The Portrait of a Lady*. Mac Pigman, as he has on previous occasions, assisted me with the original typesetting of the book, and much of the research (to call it that) was done at the Henry E. Huntington Library, the most civilized place in the world to read Victorian novels.

J.S.

Jane Austen · *Mansfield Park*

Where does Sir Thomas's wealth come from?

Edward Said's book *Culture and Imperialism*[1] was well received in the United States, but provoked some bad-tempered responses in the United Kingdom (notably in the *TLS*). The reason for the bad temper, one might suspect, was that as the imperial power principally targeted in his book's historical discussions there remained a legacy of colonists' guilt in Britain. Particular exception was taken by British commentators to Said's chapter 'Jane Austen and Empire', and its triumphant conclusion: 'Yes, Austen belonged to a slave-owning society' (p. 115). The central piece of evidence for a meaningful conjunction between the author of *Mansfield Park* and black men sweating under some sadistic overseer's whip is Sir Thomas Bertram's absence for the early stages of the novel in his estate in Antigua, during which period his unsupervised offspring put on a domestic production of Kotzebue's scandalous (for the time) play, *Lovers' Vows*.[2] Fanny Price, the waif who has been brought as a penniless young dependant to Mansfield Park, strenuously declines to participate in this godless activity. After various trials of her goodness she eventually wins the heart of the second son (and, given his elder brother's ruined health, prospective heir to the estate) Edmund, correcting in the process his wayward sense of vocation. (Edmund's 'ordination' as a clergyman was given by Austen in a cryptic remark in a letter as her novel's principal subject-matter, although there is critical dispute about just what she meant; see p. xv). Finally, after a symbolic rejection of her sordid family home at Portsmouth, Fanny is adopted by Sir Thomas as the presiding spirit of Mansfield Park. Said sees her apotheosis as an installation of world-historical significance:

> Like many other novels, *Mansfield Park* is very precisely about a series of both small and large dislocations and relocations in space that occur before, at the end of the novel, Fanny Price becomes the spiritual mistress of Mansfield Park. And that place itself is located by Austen at the centre of an arc of interests and concerns spanning the hemisphere, two major seas, and four continents. (p. 112)

Miss Fanny Price—one of the most retiring heroines in the history of English fiction—emerges transformed from Said's analysis as a pre-Victoria, empress (and oppressor) of a dominion over which the sun never sets.

According to Said: 'The Bertrams could not have been possible without the slave trade, sugar, and the colonial planter class' (p. 112). But, 'Sir Thomas's infrequent trips to Antigua as an absentee plantation owner reflect the diminishment of his class's power' (p. 113). Said is here building on two brief comments early in the novel's action. On page 20 the narrator records that Sir Thomas's circumstances 'were rendered less fair than heretofore, by some recent losses on his West India Estate, in addition to his eldest son's [Tom's] extravagance'. A little later, on page 26, Mrs Norris observes to Lady Bertram that Sir Thomas's means 'will be rather straitened, if the Antigua estate is to make such poor returns'. It is clear, however, that on her part Lady Bertram does not anticipate any serious 'diminishment' of her family's position. 'Oh! *that* will soon be settled', she tells Mrs Norris. And indeed, on his unexpected return, Sir Thomas confirms that he has been able to leave the West Indies early because 'His business in Antigua had latterly been prosperously rapid'. At the end of the novel, the Bertram estate—with Tom chastened and sober—seems on a sounder footing than ever.

Anyone attempting a historical reading should note that the period in which *Mansfield Park's* action is set (between 1805 and February 1811, when Jane Austen began writing) was not the period in which the British Empire fell, but the prelude to its extraordinary rise.[3] The year following the novel's publication, 1815 (Waterloo year), marked the beginning of imperial Britain's century. If we follow Said, this imperial achievement was a bourgeois rather than an aristocratic thing. The co-opting of middle-class Fanny Price into the previously exclusively aristocratic enclave of Mansfield Park predicts the new bourgeois energies of nineteenth-century British imperialism. The patrician absentee landlord like Sir Thomas will yield to the earnest (and essentially middle-class) district commissioner, missionary, and colonial educator (the class represented most spectacularly by the Arnolds). Fanny Price leads on, inexorably, to that wonderful apostrophe to the battalions of British 'Tom Browns' at the opening of Hughes's novel—ordinary young men and women from ordinary backgrounds, who have helped colour the bulk of the globe red.[4]

Said's insights are coolly argued and persuasive. They also supply an attractive way of teaching the novel, and will be adopted in any number of courses on post-coloniality and literature of oppression. Inevitably they will surface as orthodoxy in A level answers ('"Austen belonged to a slave-owning society": Discuss.') There are, however, a number of problems. One obvious objection is that Jane Austen seems to take the Antigua business much less seriously than does Edward Said. Like the French wars (which get only the most incidental references in

Persuasion), Austen seems to regard affairs of empire as something well over the horizon of her novel's interests. So vague is her allusion to what Sir Thomas is actually doing abroad that Said is forced into the awkward speculation, 'Sir Thomas's property in the Caribbean *would have had to be* a sugar plantation maintained by slave labour (not abolished until the 1830s)' (my italics). Not necessarily. Since Jane Austen says nothing specific on the subject, the Bertrams could have had a farm supplying produce and timber to other plantations. And if, as Said claims, Sir Thomas is doing badly, it could be because he ill-advisedly chose to raise some other crop than sugar, or because he declines to use slave labour as heartlessly as his fellow plantation owners (there is, as we shall see, some evidence for this hypothesis).

There is one rather tantalizing reference to slavery in Volume 2, Chapter 3 of *Mansfield Park*. Fanny tells Edmund that 'I love to hear my uncle talk of the West Indies. I could listen to him for an hour together' (p. 177). She continues:

> 'I do talk to him more than I used. I am sure I do. Did not you hear me ask him about the slave trade last night?'
>
> 'I did—and was in hopes the question would be followed up by others. It would have pleased your uncle to be inquired of farther.'
>
> 'And I longed to do it—but there was such a dead silence!' (p. 178)

Dead silence pretty well describes *Mansfield Park's* dealing with Antigua generally. Edward Said gets round this absence of reference by a familiar critical move. Texts, just like their readers, have their repressed memories and their 'unconscious'. Austen's not mentioning colonial exploitation betrays neurotic anxiety on the subject. In his stressing absent presences, Said is following a trail flamboyantly blazed by Warren Roberts's monograph *Jane Austen and the French Revolution* (London, 1979), a work written at the high tide of theoretic 're-reading' of classic texts. Roberts's line goes thus: as is well known, Jane Austen never mentions the French Revolution. Therefore it must be a central preoccupation, and its silent pressure can be detected at almost every point of her narratives. In *Mansfield Park*, Roberts argued for a quite specific time setting of 1805–7, when the French blockade had disastrous implications for the British sugar trade, forcing down the price to the growers from 55s. to 32s. a quintal. It is deduced that this 1805–7 crisis accounts for Sir Thomas's urgent trip to Antigua.

There is, however, nothing in the novel to confirm this historically significant date. If anything, the date markers which the narrative contains rather contradict 1805–7. There are clear references, for instance, to the *Quarterly Review* (which was not founded until 1809), Crabbe's *Tales* (not published until September 1812), and the imminent 1812 war with America. (See notes to pp. 94, 141.) These suggest a setting exactly contemporaneous with Jane Austen's writing the novel, 1811–13. On the other hand, references to *The Lay of the Last Minstrel* (published in 1805) but none of Scott's subsequent poems, could he thought to confirm a setting of 1805–7. It remains a moot point.[5] The strongest argument against an 1805–7 setting is, of course, the Crabbe reference. His *Tales* were published in September 1812 (by Hatchard, in two volumes). But, if one examines the relevant passage carefully, there is some warrant for thinking that Austen may not have had this specific 1812 publication in mind. Edmund, at the end of Volume 1, Chapter 16, asks Fanny: 'How does Lord Macartney go on?' Without waiting for answer he opens up some other volumes on the table which Fanny has apparently been reading: 'And here are Crabbe's Tales, and the Idler, at hand to relieve you, if you tire of your great book' (p. 141). The 'great book' is identified as Lord

Macartney's *Journal of the Embassy to China* (1807), which Fanny is dutifully reading since Sir Thomas is currently also on a trip abroad. It would make sense (from Edmund's ironic use of the term 'great') to assume it is a new book. And if one lower-cased the first letter of the Crabbe reference, so that it read 'and here are Crabbe's tales', it could as well refer to the much-reprinted *The Borough* or the verse stories in *Poems* (1807). Austen's manuscript, as R. W. Chapman reminds us, was not always precise on such details. There is at least sufficient uncertainty in the matter for one to consider carefully Roberts's 1805–7 hypothesis, and the elaborate superstructure he erects on it.

It remains, however, hypothetical in the extreme, and in passing one may note that Roberts's book marked a new gulf which had opened between the advanced literary critics of the academic world and the intelligent lay reader for whom, if Jane Austen or any other novelist did not mention something, it was because they did not think that something worth mentioning. *Jane Austen and the French Revolution* provoked one of the most amusing of *New Statesman* competitions, asking sportive readers of the magazine to come up with the most unlikely titles for literary critical works they could think of. The winner was the delightful: '*My Struggle*, by Martin Amis.'

It will not, of course, do to laugh Edward Said's carefully laid arguments out of court. But one can question certain aspects of his reading of *Mansfield Park*. On historical grounds one can question Said's contention that Sir Thomas's wealth comes primarily from his colonial possessions and that his social eminence in Britain is entirely dependent on revenues from Antigua. During the Napoleonic War large landholders (as Sir Thomas clearly is) made windfall fortunes from agriculture, sheep-farming, and cattle-farming. (Although, as Marilyn Butler points out, agricultural wages fell during the period, and southern England became 'a relatively depressed area', p. xiii.) Walter Scott—who had toyed with the idea of emigrating to the West Indies in 1797—discovered when he took over a farm at Ashestiel in 1805 that he could enrich himself by raising sheep, and by subletting portions of his rented agricultural land. It was the euphoria engendered by this bonanza that inspired him to go on a farm-buying spree around his 'cottage' (later a baronial mansion) at Abbotsford.[6] It led to disaster when the value of agricultural land and produce slumped in the postwar period, after 1815. If *Mansfield Park* is set at some point around 1805—13 (taking the most relaxed line on the question), Sir Thomas may conceivably have been coining it from rented and agent-managed farms on his estate. If so, Mansfield Park itself would have been the main source of his income, and would have compensated for any Caribbean shortfall.

According to Said, the fact that Fanny Price shows so little concern about what goes on Antigua is a measure of how successful the imperialist ideology was in neutralizing 'significant opposition or deterrence at home' (p. 97). This most artificial of economic arrangements—a small, northern island sucking wealth from a Caribbean island by means of workers forcibly expatriated from Africa—was rendered a fact of nature. Something so natural, in fact, that it provoked no more comment than the sun's rising in morning and setting at night. It is a beguiling argument. But it can be objected that there was indeed 'significant opposition' to the colonial exploitation of slaves in England in the early nineteenth century, and that Fanny Price—particularly as elevated by Sir Thomas's favour after his return from Antigua—would have been in the forefront of it. It is useful, in making this point, to look at an earlier critical commentary. Interest in the

Antigua dimension of *Mansfield Park* is, as it happens, of fairly long standing. The first detailed reference I have come across is by Stephen Fender (a critic specializing in immigration studies) in 1974, in a conversation recorded for a publisher of educational tapes for British sixth-formers. Fender asked:

Can it be said that Jane Austen is concerned with the 'real' social life of her time? The answer is yes. The house, Mansfield Park, no longer fills its 'ideal' role, its members no longer fulfil their functions. Tom Bertram's relationship with the land is occasional and predatory—he only comes home to hunt—and the land no longer supports the house. Its wealth is from Antigua which produced sugar which had been worked by slaves. The Wilberforce anti-slavery movement was at its height when *Mansfield Park* was written and its contemporary readers would see that the house was, in a sense, 'alienated' from its environment. Perhaps in this context it is significant that the moral inheritor of the true values of 'Mansfield Park' is Fanny Price, the outsider.[7]

Fender's point is that *Mansfield Park* is as much a novel about the English class system and its resilient modes of regeneration as it is about British imperialism. One should also note that the novel contains clear indications that Fanny Price belongs to the Clapham Sect of evangelical Christianity, which hated plays and light morality only slightly less than it loathed slavery. Her prejudices are centrally aligned with those of the sect's 'Reform of Manners' campaign. These evangelicals were mobilized politically by William Wilberforce, who allied them with the British abolitionist movement. It may well have been Wilberforce's successful bill for the abolition of slavery in 1807 which inspired Fanny's artless question to Sir Thomas about slaves. Wilberforce's bill remained a dead letter until, five years after his death in 1838, slavery in the West Indies was finally abolished. But, it is safe to say,

Fanny would have been on the side of the abolitionists from the first—as much a hater of human slavery in 1811 as she was a distruster of domestic theatricals, and from the same evangelical motives.[8]

If we take the Antigua dimension of *Mansfield Park* seriously, reading more into it than the slightness of textual references seem to warrant, it is clear that far from buttressing the crumbling imperial edifice Fanny will, once she has power over the estate, join her Clapham brethren in the abolitionist fight. Jane Austen (in 1814) may well, as Edward Said reminds us, have belonged to a slave-owning society. Fanny Price's creator died in 1817, while slave owning was still a fact of British imperial rule, whatever the Westminster law-books said. Fanny Price, we apprehend, will survive to the 1850s, before dying as Lady Bertram, surrounded with loving grandchildren and a devoted husband, now a bishop with distinctly 'low', Proudie-like tendencies. Both of them will take pride in the fact that there is no taint of slave-riches in their wealth—and that England has rid herself of the shameful practice of human bondage a full decade before the French, and no less than thirty years before the Americans. Fanny Price, and her docile husband, will certainly have done their bit to bringing this happy end about.[9]

The Oxford World's Classics *Mansfield Park* is edited by James Kinsley, with an introduction by Marilyn Butler.

Walter Scott · *Waverley*

How much English blood (if any) does Waverley spill?

Scott, in a much-quoted comment to his friend J. B. S. Morritt, expressed a low opinion of the hero of his first novel. Edward Waverley, the author of *Waverley* declared, was 'a sneaking piece of imbecility'.[1] One of the more extraordinary aspects of Waverley's imbecility is that—as far as one can make out from the account given in the narrative—he wanders through the battlefields of the great 1745 Rebellion offering as little danger to his English foe as a dormouse in a tiger's cage. Take, for instance, the highpoint in Chapter 24 of the second volume ('The Conflict'), which describes the Battle of Prestonpans in which the Scottish forces (under whose flag Waverley now fights, although he is still technically an officer of the English crown) won a famous victory, suggesting that they might indeed overrun England and restore the Stuarts to the throne. For the English military, Prestonpans was a chilling reminder of how formidable highlanders were in hand-to-hand engagement. The English commanders fondly thought that the bare-legged barbarians would be so overwhelmed by the novelty of artillery, that they would turn tail in fear when the first shells and cannon-balls exploded among them. Instead, the Scots flanked their static English foe and fell on them with cold steel. As Burton's *History of Scotland* records:

A slaughter of a frightful kind thus commenced, for the latent ferocity of the victors was roused, and grew hotter and hotter the more they pursued the bloody work. To men accustomed to the war of the musket and bayonet, the sword-cut slaughter was a restoration of the more savage-looking fields of old, which made even the victorious leaders shudder.[2]

Waverley's conduct on the field at Prestonpans is less than bloodthirsty. Observing among the mêlée an English officer 'of high rank', he is so struck by 'his tall martial figure' that he decides 'to save him from instant destruction' (p. 225), going so far as to turn the battle-axe of Dugald Mahoney, which is about to descend on Colonel Talbot's head. Waverley, who evidently has a keen eye for insignia of rank, then perceives another English colonel in trouble. 'To save this good and brave man, became the instant object of Edward's anxious exertions' (p. 226). But try as he does, again apparently impeding his own sworn comrades from their business of killing Englishmen, he can only witness the death of his former commander Colonel Gardiner and suffer his withering *et tu, Brute?* look.

What on earth, one wonders, is Waverley doing on this battlefield, scurrying around trying to save enemy *officers* from being killed? The reader is not helped by Scott's account of the battle which is remarkably patchy and vague (the event is 'well known', the narrator says by way of excuse: p. 225). Yet, we later learn, the Chevalier 'paid [Edward] many compliments on his distinguished bravery' (p. 238). After the battle Fergus informs Edward that 'your behaviour is praised by every mortal, to the skies, and the Prince is eager to thank you in person; and all our beauties of the White Rose are pulling caps for you' (p. 237). Captain Waverley is a Scottish hero—what does this mean but that he did great slaughter among the enemy? After the battle, we are told, the field is 'cumbered with carcases' (p. 230). Fergus, who is a fire-eater and a merciless warrior; would hardly praise Edward for saving his former

English comrades from destruction. 'You know how he fought' (p. 249), Rose later reminds Flora. Would that we did.

From the lustre which attaches to him after the battle, we have to assume Edward did at least a fair share of killing. But Scott's evasive narrative begs the question: did Waverley kill any Englishmen? Did his sword pierce and skewer English guts? Did he cut English throats or cave in English skulls? Did he so much as draw a drop of English blood? Soldiers on battlefields have only one mission—to kill the enemy. Either Edward Waverley is the most incompetent warrior who ever lived or—still bearing the English king's commission—he killed the English king's men.

Scott was clearly in a personal dilemma in this, all-important, aspect of his hero's military exploits. The king's commission which he himself held as a captain in the Edinburgh Light Cavalry was the most treasured possession of Scott's manhood, more dear to him by far than his being laird of Abbotsford or 'the author of *Waverley*'.[3] When, a few months after his novel's triumph, he visited Brussels and the field at Waterloo, it was in his cavalry officer's uniform that Captain Scott chose to be presented to the Tsar of Russia (it led to an unfortunate misunderstanding when the potentate assumed the novelist had been wounded in the recent battle—Scott had, in fact, been lamed from childhood by polio). Scott, the serving officer, would have been nauseated by the idea of Edward's killing fellow holders of the king's commission—it would have been a treachery deeper than Judas's. Other ranks—'erks', 'PBI' ('poor bloody infantry'), 'grunts': they have always attracted names testifying to their subhumanity—were something else altogether. What Scott intimates by highlighting Edward's stout protection of his fellow (English-Hanoverian) officers in their extremities of danger on the battlefield is that his killing (for which he had his due of fame among the Stuart ladies) was reserved for the English other ranks—those uncommissioned, unregarded, private soldiers and NCOs who have always been treated by their commissioned betters as expendable battle-fodder. In *Waverley* they are of no more account than the horses who die on the battlefield or the crows who peck at the corpses. Yes, the perceptive reader apprehends, Waverley did indeed spill English blood and a lot of it—but it was not blue blood. Had it been, his head might well have joined Fergus's on the pikes at Carlisle.

The Oxford World's Classics *Waverley* is edited by Claire Lamont.

Jane Austen · *Emma*

Apple-blossom in June?

The early nineteenth-century novelists inherited from their Gothic predecessors a sense that, where landscape was concerned, lies were more beautiful than truth and, for that reason, often preferable. In his essay on Mrs Radcliffe in *The Lives of the Novelists*, Scott notes the pervasive vagueness of her scene-painting, a quality which at its best aligns her word-drawn settings with the imaginary landscapes of Claude:

> Some artists are distinguished by precision and correctness of outline, others by the force and vividness of their colouring; and it is to the latter class that this author belongs. The landscapes of Mrs Radcliffe are far from equal in accuracy and truth to those of her contemporary, Mrs Charlotte Smith, whose sketches are so very graphical, that an artist would find little difficulty in actually painting from them. Those of Mrs Radcliffe, on the contrary, while they would supply the most noble and vigorous ideas, for producing a general effect, would leave the task of tracing a distinct and accurate outline to the imagination of the painter. As her story is usually enveloped in mystery, so there is, as it were, a haze over her landscapes, softening indeed the whole, and adding interest and dignity to particular parts, and thereby producing every effect which the author desired, but without communicating any absolutely precise or individual image to the reader. (pp. 118–19)[1]

For all the realism of his historical analysis and characterization, Scott often found a similar 'haze' very useful in his own higher-flying landscape descriptions. It was pointed out to him when embarking on *Anne of Geierstein*

(1829) that it might be a handicap never to have visited the Swiss Alps, where the action is set. Nonsense, Scott replied, he had seen the paintings of Salvator Rosa, and that would do very well, thank-you.[2] Radcliffian haze was also very useful to Scott in what remains the most famous anomaly in his fiction, the 'reversed sunset' in *The Antiquary* (1816). In an early big scene in that novel, Sir Arthur Wardour and his daughter Isabella are trapped between the onrushing tide and unscaleable cliffs. The location is identifiably Newport-on-Tay (called in the novel 'Fairport'), near Dundee, on the east coast of Scotland. Scott highlights the scene by having it occur while the great disc of the sun sinks into the North Sea—a lurid panorama on which two paragraphs of fine writing is lavished.

The problem is, of course, that in our cosmos the sun does not sink in the east, it sinks in the west, in the Irish Sea. Given the haste with which he wrote his novel it is not surprising, perhaps, that Scott should have perpetrated the error. What is surprising is that he should have retained it in his 1829 revised edition of *The Antiquary.* The mistake was certainly pointed out to him. Evidently he felt that where land and seascapes were concerned, the novelist's artistic licence extended to changing the course of the planets through the heavens. Novelists later in the century were more fastidious. Rider Haggard, for instance, rewrote large sections of *King Solomon's Mines* in order to correct an error about the eclipse of the sun which is so technical as to be beyond all but the most astronomically expert readers.[3] Haggard mistakenly had the solar eclipse occur while the moon was full. In all editions of *King Solomon's Mines* after the '37th thousand' he changed it to a lunar eclipse.

This fetishism about scenic detail develops in the 1830s and 1840s. It may well have coincided with more sophistication about the authenticity of theatrical sets, a greater awareness of what foreign parts looked like with the growth of the British tourism industry, and the diffusion of encyclopaedias among the novel-reading classes. Captain Frederick Marryat wrote *Masterman Ready, or the Wreck of the Pacific* (1841) specifically to correct the travesty of life on a South Seas desert island perpetrated by Johann Wyss's *The Swiss Family Robinson* (1812, 1826). Marryat, who as a sailor had felt the brine of the seven seas on his cheek, was appalled by such freaks of nature as flying penguins and man-eating boa-constrictors.[4]

Jane Austen's most lamentable landscape-painting error occurs in the Donwell picnic scene in *Emma*. The date of the picnic is given to us very precisely 'It was now the middle of June, and the weather fine', we are told on page 319. And again, on page 323, the excursion is described as taking place 'under a bright mid-day sun, at almost Midsummer' (i.e. around 21 June). Strawberries are in prospect, which confirms the June date. During the course of the picnic, Austen indulges (unusually for her) in an extended passage describing a distant view—specifically Abbey-Mill Farm, which lies some half-a-mile distant, 'with meadows in front, and the river making a close and handsome curve around it'. The narrative continues, weaving the idyllic view into Emma's tireless matchmaking activities:

> It was a sweet view—sweet to the eye and the mind. English verdure, English culture, English comfort, seen under a sun bright, without being oppressive.
>
> In this walk Emma and Mr Weston found all the others assembled; and towards this view she immediately perceived Mr Knightley and Harriet distinct from the rest, quietly leading the way. Mr Knightley and Harriet!—It was an odd tête-à-tête; but she was glad to see it.—There had been a time when he would have scorned her as a companion, and turned from her with little

ceremony. Now they seemed in pleasant conversation. There had been a time also when Emma would have been sorry to see Harriet in a spot so favourable for the Abbey-Mill Farm; but now she feared it not. It might be safely viewed with all its appendages of prosperity and beauty, its rich pastures, spreading flocks, orchard in blossom, and light column of smoke ascending. (pp. 325–6)

James Kinsley offers a note to 'in blossom':

The anomaly of an orchard blossoming in the strawberry season was noticed by some of the novel's first readers. Jane Austen's niece Caroline wrote to a friend as follows: 'There is a tradition in the family respecting the apple-blossom as seen from Donwell Abbey on the occasion of the strawberry party and it runs thus—That the first time my uncle . . . saw his sister after the publication of *Emma* he said, "Jane, I wish you would tell me where you get those apple-trees of yours that come into bloom in July." In truth she did make a mistake—there is no denying it—and she was speedily apprised of it by her brother—but I suppose it was not thought of sufficient consequence to call for correction in a later edition.' (p. 444)

One could defend the anachronistic apple-blossom in the same way that one defends the anastronomical sunset in that other novel of 1816, *The Antiquary*. Both represent a hangover from the free-and-easy ways of the Gothic novel of the 1790s when such liberties could be taken with artistic impunity. But this is not entirely satisfactory with the author of *Northanger Abbey*, a novel which hilariously castigates Gothic fiction's offences against common sense. And, as R. W. Chapman notes (apropos of the apple blossom), such mistakes are 'very rare' in Miss Austen's fiction.[5]

It was evidently assumed by Jane Austen's family that no correction was made because the error was 'not thought of sufficient consequence'. This is unlikely; elsewhere one can find Jane Austen going to some length to authenticate detail in her fiction (she put herself to trouble, for instance, to verify details as to whether there was a governor's house in Gibraltar, for *Mansfield Park*).

If the 'apple-blossom in June' error were pointed out to her, why then did Jane Austen not change it? 'Orchards in leaf' would have been an economical means of doing so, requiring no major resetting of type. One explanation is that she did not have time—some eighteen months after the publication of *Emma* Jane Austen died, in July 1817. A more appealing explanation is that it is not an error at all. It was not changed because the author did not believe it was wrong. In order to make this second case one should note that there is not one 'error' in the description (blossom in June), but two, and possibly three. Surely, on a sweltering afternoon in June, there would not be smoke rising from the chimney of Abbey-Mill Farm? Why have a fire? And if one were needed for the baking of bread, or the heating of water in a copper for the weekly wash, the boiler would surely be lit before dawn, and extinguished by mid-morning, so as not to make the kitchen (which would also be the family's dining-room) unbearably hot. The reference to the ascending smoke would seem to be more appropriate to late autumn. And the reference to 'spreading flocks' would more plausibly refer to the lambing season, in early spring, when flocks enlarge dramatically. It will help at this point to quote the relevant part of the passage again: 'It might be safely viewed with all its appendages of prosperity and beauty, its rich pastures, spreading flocks, orchard in blossom, and light column of smoke ascending.' What this would seem to mean is that now Harriet is so effectively separated from Mr Robert Martin, the occupant of Abbey-Mill Farm, she is immune to its varying attractions over the course of the year—whether in spring, early summer, midsummer or autumn. What Austen offers us in this sentence is not Radcliffian haze, but a precise depiction, in the form of a miniature montage, of the turning seasons.[6] Months may come and months may go, but Harriet will not again succumb to a mere farmer.

The Oxford World's Classics *Emma* is edited by James Kinsley with an introduction by Terry Castle.

Walter Scott · *The Heart of Midlothian*

Effie Deans's phantom pregnancy

The wonderful plot of *The Heart of Midlothian*—Jeanie Deans's refusing to perjure herself in court to save her sister's life, and her tramp down to London to beg mercy from the queen—originate in the misfortune of Effie's pregnancy. Yet that misfortune, closely examined, is an extremely problematic thing. Certain improbabilities in it force us to assume either, (1) that Scott found himself trapped in a narrative difficulty which he could not easily write himself out of, or (2) that there is more to the episode than meets the casual reader's eye.

I prefer the second of these assumptions, although in support of the first it should be said that the law by which Effie is condemned contains some very dubious propositions. Scott outlines the 1690 law (which was repealed in the early nineteenth century) in a note to Chapter 15. A woman was liable to execution for infanticide on the circumstantial grounds 'that she should have concealed her situation during the whole period of pregnancy; that she should not have called for help at her delivery; and that, combined with these grounds of suspicion, the child should be either found dead or be altogether missing' (p. 528). What is unlikely is that a woman, in any normal social situation, should be able to disguise her altered physical shape in the last months of pregnancy, or that she should be able to deliver her own child without the assistance of a midwife. Common sense suggests that the ordinance must have been most effective, not against infanticide, but abortion. A woman might well conceal her condition for four or five months and procure an abortion, at the actual climax of which the abortionist might be prudently absent. Women likely to fall foul of the law would not anyway be reliable as to when menstruation stopped, and 'the whole period of pregnancy' would be very hard to establish.

In Chapter 10 of *The Heart of Midlothian* we are told, in a rapid summary way, how Effie is sent to work in the Saddletrees' shop, by the Tolbooth Kirk, in the High Street. Her half-sister (who is some years older, less beautiful, and more religious by nature) is pleased, since there have been signs of lightness in Effie's conduct—most worryingly a propensity to dancing and unidentified 'idle acquaintances' which the young girl has formed around St Leonard's Crags, at the south side of the city, where her father keeps his small dairy herd.

At first things go well with Effie, who lodges with Mr and Mrs Saddletree (something of a termagant). The young woman does her work cheerfully and well. They are relatives, and kind employers. But:

> Ere many months had passed, Effie became almost wedded to her duties, though she no longer discharged them with the laughing cheek and light step, which at first had attracted every customer. Her mistress sometimes observed her in tears, but they were signs of secret sorrow, which she concealed as often as she saw them attract notice. (pp. 103–4)

Her 'disfigured shape, loose dress, and pale cheeks' inevitably attract the 'malicious curiosity' and the 'degrading pity' of neighbours and fellow-servants. But she confesses nothing, returning all taunts with bitter sarcasm. Why, one wonders, do not her master and mistress (who are living alongside her) note their kinswoman's alteration as readily as casual customers in the shop? Bartoline Saddletree, we are told, is 'too dull at drawing inferences from the occurrences of common life'. Scott concedes that

Effie's changed shape and demeanour 'could not have escaped the matronly eye of Mrs Saddletree, but she was chiefly confined by indisposition to her bedroom for a considerable time during the latter part of Effie's service' (p. 104). What this indisposition is, we are not told. But it would seem to have kept her in total quarantine, even from talkative servants (who would surely have observed Effie's interestingly changed shape). Mrs Saddletree is lively enough in subsequent sections of the narrative, suggesting a wonderfully quick recovery.

A more troubling question is why Effie's pregnancy escaped the notice of her half-sister. Jeanie, as a cow-feeder and an older woman (old enough, indeed, to remember Effie's birth) would certainly have known what a pregnant young girl looked like. Scott, rather fuzzily, suggests they never met. Mrs Saddletree's illness gave Effie a pretext for never coming to St Leonard's Crags. And 'Jeanie was so much occupied, during the same period, with the concerns of her father's household, that she had rarely found leisure for a walk into the city, and a brief and hurried visit to her sister' (pp. 104–5).

There are two problems here. The first is that—despite what readers who do not know Edinburgh may be led to think—the Deans's farm at St Leonard's Crags, on the common land beneath Arthur's Seat—is about twelve minutes walk from the High Street. For a healthy young body like Jeanie Deans (who would be bringing her produce to the High Street anyway), the excuse that she was too busy to walk a mile to visit her sister is highly unconvincing. Moreover, under oath in court Jeanie testifies that she did indeed visit her sister during the latter months of her pregnancy, noticed that she looked unwell, and questioned her as to 'what ailed her' (p. 231). No answer was given by the unfortunate young woman. None the less, when Effie returns, minus baby, to St Leonard's

Crags, Jeanie immediately deduces (without being told) what the problem is, even though Effie's shape is now more like its old self. Her half-sister has been 'ruined', Jeanie realizes—just by looking at her. This percipience, following on earlier impercipience, is very odd.

The Saddletrees' ignorance of Effie's swollen belly and desperately changed appearance is barely credible. But one would have liked a little more detail about the strange illness that kept the lady of the house (an inveterate busybody) so completely out of the affairs of her household for so many months. What is not believable is that Jeanie, having observed Effie's condition, did not make the obvious deduction, based on her knowledge of what might happen to young girls with light morals in the city. There are two possible explanations: (1) Jeanie saw, and suspected, but said nothing—in the hope that she was wrong or that some act of God might put all right; (2) that, out of anger against her sister, Jeanie deliberately kept out of Effie's way, thus leaving her exposed to danger. Guilt at having abandoned Effie at this most precarious moment of her existence may be thought to have strengthened Jeanie's determination to walk, barefoot, like some medieval self-flagellant, the length of the kingdom to win a reprieve.

The Oxford World's Classics *The Heart of Midlothian* is edited by Claire Lamont.

Mary Shelley · *Frankenstein*

How does Victor make his monsters?

Answering from one's visual recollections of the films, the answer to the question in the title is easy. Colin Clive (in James Whale's and Universal Pictures' 1931 modern-dress classic), Peter Cushing (in a string of Hammer-produced, Terence Fisher-directed gothics from 1957 onwards), James Mason (in the 1973 TV-movie, scripted by Christopher Isherwood), Gene Wilder (in Mel Brooks's hilarious 1974 spoof *Young Frankenstein*), Kenneth Branagh (in the 1994 blockbuster, *Mary Shelley's Frankenstein*), and even the heroes in such delightfully zany send-ups as *Frankenhooker* (a story based on monstrous exploding prostitutes) and *Frankenstein, the College Years* are—all of them—half surgeons of genius, half mad scientists. In this dual capacity these cinematic Frankensteins dispatch goblin-like henchmen to scour the local graveyards, gibbets, morgues, and anatomy schools for human parts. This anatomical junk is assembled by the surgeon's art like some carnal jigsaw ('can we use this brain?'), and the final inert, composite cadaver is raised on a platform to be struck—like Benjamin Franklin's kite—by lightning, harnessed into life-giving energy by the scientist's brilliant technology.

This episode has been repeated so often on the screen as to be folkloric. It is, however, wholly unfaithful to what Mary Shelley wrote and published in 1818.[1] The spare-parts, galvano-animated Frankenstein's monster derives from a number of sources well outside the source narrative. Whale's mad scientist—which has been formative—

is an elision of Frankenstein and a range of popular misconceptions about the wild-haired Nobel Prizewinning German with the same-sounding name, Albert Einstein. Hence the bizarre fact that, in the classic 1931 film, Shelley's early-nineteenth-century tale is transposed into an alien, early-twentieth-century setting. Shelley's Frankenstein is no 'scientist', whether mad or sane, but an Enlightenment *philosophe.* As the *OED* records, the word 'scientist' was not even coined until 1840.

The main distortion of Shelley's original conception of Frankenstein can be laid at the door of the novelist herself. Writing in her 1831 preface ('to the Third Edition'), the author muses about how the great experiment *might* have been achieved: 'Perhaps a corpse would be reanimated; galvanism had given token of such things: perhaps the component parts of a creature might be manufactured, brought together, and endued with vital warmth.' Shelley enlarges on this idea, picturing some 'pale student of unhallowed arts kneeling beside the thing he had put together. I saw the hideous phantasm of a man stretched out, and then, on the working of some powerful engine, show signs of life, and stir with an uneasy, half-vital motion' (p. 196).

This scenario is given precise detail (fleshed out, one might say) by graphic accounts of Percy Shelley at Oxford University showing off a small model dynamo to his friends. According to Thomas Jefferson Hogg, while they were both students at Oxford, Shelley:

> proceeded, with much eagerness and enthusiasm, to show me the various instruments, especially the electrical apparatus: turning round the handle very rapidly so that the fierce, crackling sparks flew forth; and presently standing on the stool with glass feet, he begged me to work the machine until he was filled with the fluid [i.e. electrical current], so that his long wild locks bristled and stood on end. Afterwards he charged a powerful battery of

several large jars; labouring with vast energy and discoursing with increasing vehemence of the marvellous powers of electricity, of thunder and lightning; describing an electrical kite that he had made at home, and projecting another and an enormous one, or rather a combination of many kites, that would draw down from the sky an immense volume of electricity, the whole ammunition of a mighty thunderstorm; and this being directed to some point would there produce the most stupendous results.[2]

As Maurice Hindle observes, in his Penguin Classics edition *of Frankenstein*, 'it is almost as if Shelley were providing film-makers of the future with that "thunder and sparks" image of electrical reanimation of the Creature which has become so standard a feature of Frankenstein films'.[3] Film-makers, as we shall see, were certainly grateful for the cue. But before moving-films arrived, *Frankenstein* had established itself as an immensely popular melodrama in the theatre in the nineteenth century, and had seeped from the stage into proverbial currency (it was, for instance, commonplace for Victorian cartoonists to depict rampaging working-class or Irish mobs as 'Frankensteins'). There were, as Steven Forry calculates, over ninety different dramatizations of Mary Shelley's novel in the nineteenth century—almost as many, in fact, as there have been films in the twentieth century.[4] All these nineteenth-century theatrical versions, lacking a complex stage machinery with which to simulate 'creation', presented a monster manufactured by alchemy or some *elixir vitae.*

The possibilities of Shelley's story were quickly apprehended by the fledgling American movie industry of the early twentieth century. The first serious film version of *Frankenstein* was produced by Thomas A. Edison's company, in 1910.[5] Although Edison was the arch-apostle of 'electricity', this pioneer film of *Frankenstein* retained the alchemically generated monster of nineteenth-century melodrama. It was actually a 1915 Broadway stage adaptation, called *The Last Laugh*, which picked up on Mary Shelley's afterthought about galvanism by introducing electrical machinery into the creation scene.[6] And this motif was taken over and enlarged in the 1931 Universal Pictures production, with its massive laboratory set, dominated by huge electro-galvanic generators, designed by Kenneth Strickfadden. As has been observed, there are clear analogies between Frankenstein's laboratory and the cavernous studio in which the film was shot (also powered by electrical generators). But James Whale was more directly influenced by an elaborately staged scene in Fritz Lang's dystopian epic *Metropolis* (1926), where a robot is electrically activated. Whale carried over into his lumbering Karloff-creation of Frankenstein's 'monster' a robotic clumsiness and total inarticulacy (not to mention a metallic bolt attaching its head to its body) which is wholly alien to Shelley's highly literate and athletically lithe monster, who trains himself in human emotion by reading Goethe's *Sorrows of Young Werther* and is an expert rock-climber.

The electro-robotic-Karloffian stereotype of Frankenstein dominated popular reproductions until virtually the present day. By the 1990s, however, bio-genetics had taken over from physics as the cutting-edge science. What Relativity was in the 1920s, the Genome Project now is. As Kenneth Branagh notes in his afterword to the 'novelization' of *Mary Shelley's Frankenstein*: 'we hope audiences [of the 1994 film] today may find parallels with Victor today in some amazing scientist who might be an inch away from curing AIDS or cancer and needs to make some difficult decisions.'[7]

This is clearly a new turn in the *Frankenstein* industry, and reflects the state of things at the end of the twentieth century. What Branagh's film shares with its predecessors, however, is a morbid fascination with the scientific details of the creation process, even if it disagrees on the precise scientific bases of that process. In the novelization's text (reflecting the space devoted to the episode in the film), no less than three chapters are expended on mechanical description of the great life-giving experiment. We are taken painstakingly through the manufacture of the gigantic copper sarcophagus in which the embryonic monster is suspended, the collection of body parts (including the *de rigeur* genius's brain), the pooling of many buckets of amniotic fluid into a sufficient placenta, the construction of the electrical generating apparatus (with a tankful of electric eels as back-up—a fine surreal touch).

What happens in Mary Shelley's source narrative is very different from what we recall from film treatments of her novel (even that latest version, which proclaims itself *Mary Shelley's*). Victor Frankenstein's initial researches are into alchemy via the ancient wisdom of Cornelius Agrippa, Albertus Magnus, and Paracelsus (p. 23). He progresses beyond the limits of this sterile and antique book-learning on witnessing the effects of a thunderstorm:

> As I stood at the door, on a sudden I beheld a stream of fire issue from an old and beautiful oak, which stood about twenty yards from our house; and so soon as the dazzling light vanished, the oak had disappeared, and nothing remained but a blasted stump . . . I never beheld any thing so utterly destroyed. (p. 24)

This demonstration of the awesome power of lightning turns Victor's attention away from the learning of the past towards the contemplation of 'electricity and galvanism' and the study of mathematics.[8] He subsequently leaves the schools of Geneva for those of Ingolstadt where he is exposed to the scientific puritanism of Monsieur Krempe, 'professor of natural philosophy'. Under Krempe and his colleague Monsieur Waldman, Victor principally studies chemistry. Through the revelations of modern chemical science, the young man has a blinding vision of how the old dreams of the alchemists can at last be realized. He duly moves on into biology, where he investigates 'the principle of life' (p. 33). Victor now turns his mind to the mysterious processes of decay and degeneration in animal tissue. His arduous studies bear fruit: 'After days and nights of incredible labour and fatigue, I succeeded in discovering the cause of generation and life; nay, more, I became myself capable of bestowing animation upon lifeless matter' (p. 34).

The stage is now set for the novel's great scene—creation. But, Victor warns his correspondent, he will be 'reserved upon the subject', and reserved he certainly is. Evidently, he collects and arranges various materials for his experiment: but what precisely these materials might be is not divulged. There follows a piece of magnificent Gothic fuzz:

> One secret which I alone possessed was the hope to which I had dedicated myself; and the moon gazed on my midnight labours, while, with unrelaxed and breathless eagerness, I pursued nature to her hiding places. Who shall conceive the horrors of my secret toil, as I dabbled among the unhallowed damps of the grave, or tortured the living animal to animate the lifeless clay? My limbs now tremble, and my eyes swim with the remembrance; but then a resistless, and almost frantic impulse, urged me forward; I seemed to have lost all soul or sensation but for this one pursuit. It was indeed but a passing trance, that only made me feel with renewed acuteness so soon as, the unnatural stimulus ceasing to operate, I had returned to my old habits. I collected bones from charnel houses; and disturbed, with profane fingers, the tremendous secrets of the human frame. In a solitary chamber, or rather cell, at the top of the house, and separated from all the other apartments by a gallery and staircase, I kept my workshop of filthy creation; my eyeballs were starting from their sockets in attending to the details of my employment. The dissecting room and the

slaughter-house furnished many of my materials; and often did my human nature turn with loathing from my occupation, whilst, still urged on by an eagerness which perpetually increased, I brought my work near to a conclusion. (pp. 36–7)

A number of interesting features stand out from the swirling rhetoric. Clearly Victor Frankenstein has undertaken field-work which would get him disbarred from most self-respecting medical associations: in morgues ('charnel houses'), graveyards, the vivisectionist's bench, the anatomy theatre, and—interestingly—the animal abattoir ('slaughter-house'). It remains unclear whether his motive has been research into primal tissue, or the kleptomaniac filching of limbs and organs with which Fritz's midnight forays in the films have made us familiar.[9] The most urgent feature of Victor's description is the emetic disgust and the narrator's irresistible urge to avert his eyes from his 'workshop of filthy creation'.

Finally, on the fateful, dreary night in November (five months later), the monster comes to life—not under the stimulus of some immense bolt of meteorological energy, but simply by opening his eyes and shuddering, like any other newborn human. A veritable convulsion of nauseated disgust follows Victor's first sight of his animated 'creation':

How can I describe my emotions at this catastrophe, or how delineate the wretch whom with such infinite pains and care I had endeavoured to form? His limbs were in proportion, and I had selected his features as beautiful. Beautiful!—Great God! His yellow skin scarcely covered the work of muscles and arteries beneath; his hair was of a lustrous black, and flowing; his teeth of a pearly whiteness; but these luxuriances only formed a more horrid contrast with his watery eyes, that seemed almost of the same colour as the dun white sockets in which they were set, his shrivelled complexion, and straight black lips. (p. 39)

The mysterious animating operation is repeated later in the narrative, when Victor is induced to 'create' a mate for his monster. He does so (surreally enough) in the Orkneys' wilderness. There are, apparently, no graveyards, anatomy theatres, or slaughterhouses for him to prey on in this northern wasteland. The island has only three huts, into one of which Victor moves himself and his laboratory equipment. Again we are told it is 'a filthy process', and virtually nothing more other than that the operation is even more loathsome than before:

> During my first experiment, a kind of enthusiastic frenzy had blinded me to the horror of my employment; my mind was intently fixed on the sequel of my labour, and my eyes were shut to the horror of my proceedings. But now I went to it in cold blood, and my heart often sickened at the work of my hands. (p. 137)

As far as one can make out, the 1818 text, stripped of the 1831 preface, gives no warrant for the surgery, transplants, and electrical apparatus of cinematic folklore.[10] In literary terms the creation of Victor's monster is closer to the mythical clay beast of Judaic folklore, the Golem, traditionally animated by charms (or *shems*) uttered by profane medieval rabbis. But the key to understanding what is implied in Mary Shelley's genesis passages is the physical, eye-averting revulsion—far in excess of what is actually described. It is a reflex and a rhetoric associated traditionally in Anglo-Saxon cultural discourse with two activities: sexual intercourse (and its variant, self-abuse), and childbirth (and its variant, abortion). Both of these are conventionally 'filthy'.

Mary, unlike Percy, did not have the advantage of an Oxford education where she might acquaint herself with the latest discoveries in Natural Philosophy. But she did know about making babies—too much, as it happened. The author of *Frankenstein* had good warrant for mixed feelings about sex and the 'labour' of childbirth. A strikingly original analysis of the novel along these lines was outlined by Ellen Moers in 1974. Moers read the 'creation' episode not as some alchemical transmutation, nor as a scientific experiment, but as an expression of the 'trauma of the afterbirth'. Moers traced the experiences which, over Mary's short lifetime, might have conduced to a jaundiced view of motherhood.[11] Mary Wollstonecraft Godwin was born on 30 August 1797, five months after her parents' marriage (as the Creature is born, after a mysterious five months' gestation). Mary's mother, Mary Wollstonecraft, died ten days after Mary's birth in August 1797, of puerperal fever. Mary, in a sense, killed the woman who gave her life—or, as the young child may have thought, her mother could not bear to live, having looked on her child. Barely past puberty, Mary eloped to France with Percy Shelley in July 1814. Sexual intercourse began early for her. And her devirgination was unhallowed. In February 1815 Mary (aged 17) gave premature birth to a daughter who died a few days later. In January 1816 she gave birth to a son, William. Mary and Shelley did not marry until December 1816 (a few days after the suicide of Percy's first wife, Harriet, who was pregnant by another man). While she was completing *Frankenstein* in May 1817, Mary was pregnant with her third child, Clara, who was born in September. She was just 20.

Despite her free-thinking, rationalist background, Mary may well have been ashamed of her irregular and busy sexual career. Her first unhappy experience of motherhood—which must surely have occurred before she was emotionally ready for it—was horrific. Mary's distaste for the business of procreation spills over, Moers plausibly argues, into the plot of her novel. One can enlarge on this insight, and see the trauma extending beyond the 'afterbirth' into disgust at the whole 'filthy' mechanics of sex.

Reading between the lines, the strong implication is that Victor creates his monster not by surgical manufacture, but by a process analogous to fertilization and *in vitro* culture. The initial work 'of his hands' which Victor refers to is, presumably, masturbatory, The resulting seed is mixed with a tissue, or soup composed of various tissues. The mixture is grown *ex utero* until, like the child cut from the umbilical cord, it is released into life. Victor Frankenstein, that is, is less the mad scientist than the reluctant parent, or semen donor. He does not make his monster, as one might manufacture a robot—he gives birth to him, as one might to an unwanted child, the sight of whom fills one with disgust. His revulsion for his 'Creation' may well, as Moers suggests, reflect aspects of Mary's own postnatal depressions and her distaste for the nasty business by which babies are made. The principal warrant for this speculation is, of course, the flavour of the rhetoric. Had Victor Frankenstein been the scientist of later film fame, his creation would have been 'blasphemous', 'vile', 'outrageous', or 'an offence against nature'. It would not have been, as Shelley repeatedly labels it, 'filthy'.

Over the last century-and-a-half one can see an interesting evolution in answers to the 'making Frankenstein' question. The early stage versions, because of the limitations of theatre machinery, fell back on the standby of the 'elixir of life' device (flasks are easier to come by as props than the paraphernalia of laboratories). The electrico-robotic Frankenstein comes in with the growth of the movie industry in the 1930s, and the Hollywood studio's no cost-spared attitude to special effects. In the 1970s, as a consequence of feminist rereadings of classic literary texts, a 'gynocritical' explanation is asserted by academics, and *Frankenstein* becomes the central work in the formation of a canon of feminist literary criticism.[12]

Kenneth Branagh's latest reproduction awkwardly combines in its filmic imagery the masculinist 'electric' and the feminist 'obstetric' interpretations. The monster, 1994-style, is gestated in a gigantic copper womb, suspended in gallons of amniotic fluid. Like Venus, he is born from the waves. But, come delivery time, he is animated by the traditional phallic jolt of electricity. Branagh's conciliatory imagery is, however, doomed by contradictions inherent in the source text. Victor is ineradicably male and, like bright middle-class males of the early nineteenth century, he has the privilege of a university education—something entirely denied young women of the period, however bright. Given his sex, Victor can 'invent' but he cannot plausibly 'mother' his monster. Yet many of the emotions ascribed to him in the highly charged description of his creature's creation are clearly maternal—and seem to be directed as shared experience to other women. The dilemma this raises can only be solved by a science-fiction scenario as daring as Shelley's in her novel. Were the author to be resurrected to rewrite her novel today, she might well devise a 'Victoria Frankenstein, Ph.D (Biology, MIT)'. Failing that, the reader and film-goer can look forward to many other contorted answers to the question: 'How does Victor make his monster?'

There are two Oxford World's Classics editions of *Frankenstein*. That by James Kinsley and M. K. Joseph reproduces the 1831 text. *Frankenstein (1818 Text)* is edited by Marilyn Butler, and reproduces the first version of the novel.

Charles Dickens · *Oliver Twist*

Is Oliver dreaming?

The mysterious apparition of Fagin and Monks at the window outside the room where Oliver is dozing in the supposed safety of his country retreat with the Maylie family furnishes one of Cruikshank's memorable illustrations to *Oliver Twist:*

Monks and the Jew

The circumstances surrounding this episode have been much worried over by commentators on the novel. The convalescent Oliver is described as being in his own little room, on the ground-floor; at the back of the house. The situation is Edenic:

It was quite a cottage-room, with a lattice-window: around which were clusters of jessamine and honeysuckle, that crept over the casement, and filled the place with their delicious perfume. It looked into a garden, whence a wicket-gate opened into a small paddock; all beyond was fine meadow-land and wood. There was no other dwelling near, in that direction; and the prospect it commanded was very extensive.

One beautiful evening, when the first shades of twilight were beginning to settle upon the earth, Oliver sat at this window, intent upon his books. He had been pouring over them for some time; and as the day had been uncommonly sultry, and he had exerted himself a great deal; it is no disparagement to the authors: whoever they may have been: to say, that gradually and by slow degrees, he fell asleep.

There is a kind of sleep that steals upon us sometimes, which, while it holds the body prisoner, does not free the mind from a sense of things about it, and enable it to ramble at its pleasure. So far as an overpowering heaviness, a prostration of strength, and an utter inability to control our thoughts or power of motion, can be called sleep, this is it; and yet we have a consciousness of all that is going on about us; and if we dream at such a time, words which are really spoken, or sounds which really exist at the moment, accommodate themselves with surprising readiness to our visions, until reality and imagination become so strangely blended that it is afterwards almost a matter of impossibility to separate the two. Nor is this, the most striking phenomenon incidental to such a state. It is an undoubted fact, that although our senses of touch and sight be for the time dead, yet our sleeping thoughts, and the visionary scenes that pass before us, will be influenced, and materially influenced, by the *mere silent presence* of some external object: which may not have been near us when we closed our eyes: and of whose vicinity we have had no waking consciousness.

Oliver knew, perfectly well, that he was in his own little room; that his books were lying on the table before him; and that the sweet air was stirring among the creeping plants outside. And yet he was asleep. Suddenly, the scene changed; the air became close and confined; and he thought, with a glow of terror, that he was in the Jew's house again. There sat the hideous old man, in his accustomed corner: pointing at him: and whispering to another man, with his face averted, who sat heside him.

'Hush, my dear!' he thought he heard the Jew say; 'it is he, sure enough. Come away'

'He!' the other man seemed to answer; 'could I mistake him, think you? If a crowd of devils were to put themselves into his exact shape, and he stood amongst them, there is something that would tell me how to point him out. If you buried him fifty feet deep, and took me across his grave, I should know, if there wasn't a mark above it, that he lay buried there. I should!'

The man seemed to say this, with such dreadful hatred, that Oliver awoke with the fear, and started up.

Good Heaven! what was that, which sent the blood tingling to his heart, and deprived him of his voice, and of power to move! There—there—at the window; close before him; so close, that he could have almost touched him before he started back: with his eyes peering into the room, and meeting his: there stood the Jew! And beside him, white with rage, or fear, or both, were the scowling features of the very man who had accosted him at the inn-yard.

It was but an instant, a glance, a flash, before his eyes; and they were gone. But they had recognised him, and he them; and their look was as firmly impressed upon his memory, as if it had been deeply carved in stone, and set before him from his birth. He stood transfixed for a moment; and then, leaping from the window into the garden, called loudly for help. (pp. 271–2)

Chapter 34 breaks off at this point. The next chapter opens with a general alarm at Oliver's cry ('The Jew! The Jew!').[1] Oliver points out the 'course the men had taken'.

But, despite everyone's vigorous efforts, 'the search was all in vain. There were not even the traces of recent footsteps, to be seen' (p. 275). The physical improbability of the two men having been in the area is pondered:

> They stood, now, on the summit of a little hill, commanding the open fields in every direction for three or four miles. There was the village in the hollow on the left; but, in order to gain that, after pursuing the track Oliver had pointed out, the men must have made a circuit of open ground, which it was impossible they could have accomplished in so short a time. A thick wood skirted the meadow-land in another direction; but they could not have gained that covert for the same reason. (p. 275)

But when Harry Maylie tells Oliver 'it must have been a dream', the boy protests: 'Oh no . . . I saw him too plainly for that. I saw them both, as plainly as I see you now' (p. 275). Inquiries are pursued, servants are dispatched to ask questions at all the ale-houses in the area, but nothing is turned up. Monks and Fagin have not been seen by another human eye in the neighbourhood, going or coming, although the appearance of two such low-life aliens would surely have attracted the notice of suspicious locals (as Monks immediately attracted Oliver's attention when he earlier ran into him at the nearby market town). Nor is the fact that the two men were actually at Oliver's window confirmed later in the story. Has Oliver imagined the whole thing? Was it a dream? By commissioning an illustration of the scene by Cruikshank, Dickens seems to support Oliver's insisted declaration—that it visibly and actually happened. The men *were* there at the window. Moreover, the conversation (particularly Monks's characteristically melodramatic expressions of hatred) rings very true in the reader's ear.

But, if one accepts the actuality of what Oliver saw, three problems follow: (1) How did Fagin and Monks discover where Oliver was staying? (2) How did the two men, neither of whom is notably agile, disappear so suddenly – before Oliver, who *is* agile and has jumped out of the window, can see where they have made off to? (3) Why did the interlopers leave no physical trace of their presence?

A number of explanations have been put forward. That Dickens was less careful in writing the novel than we are in reading it is the most primitive. Humphry House, in his introduction to the 1949 Oxford Illustrated edition of *Oliver Twist*, notes that Dickens wrote and published the work in a huge hurry, and that it 'was Dickens's first attempt at a novel proper. The sequence of the external events which befall Oliver [are] complicated and careless.' John Bayley elsewhere notes that Dickens repeats in this window scene material which is to be found earlier in the novel, at the beginning of Chapter 9 ('There is a drowsy state, between sleeping and waking . . .', see p. 64). This recycling of material would support the view that Dickens was under severe pressure.[2] Another primitive explanation is that in the window episode Dickens is resorting to the crude tricks of the Gothic ghost story. There is, for instance, a parallel instance earlier when, as he conspires at midnight with Fagin, Monks sees 'the shadow of a woman, in a cloak and bonnet, pass along the wainscot like a breath!' (p. 206). The two men search the whole house and find nothing—the woman, we apprehend, is the wraith of Oliver's mother, his protective angel.

Many critics prefer more ingenious readings. Steven Marcus, for example, in *Dickens: From Pickwick to Dombey* (London, 1965) examines the scene and its 'hypnagogic' references for Freudian clues that can be tracked back to Dickens's primal experiences in the blacking factory. J. Hillis Miller reads the scene for its demonstration of 'the total insecurity of Oliver's precarious happy state'. The vision of evil at the window is proof that his 'past is permanently part of him'. The absence of prints suggests that Fagin is to be equated with the similarly lightfooted devil (p. 73).[3]

Colin Williamson, reviewing these and other hypotheses, offers what he calls a more 'mundane' explanation.[4] He advocates reading *Oliver Twist* as one would a detective story or crime thriller. Williamson notes as significant a perplexing earlier episode in Chapter 32 in which Oliver is travelling to London with Mr Losberne to find the house of Mr Brownlow (Oliver can recall the street and the general aspect of the building, but not the number). Suddenly, at Chertsey Bridge, the excited boy turns very pale when he 'recognises' another house—that in which he and Sikes's gang hid before the attempted burglary (p. 249). They stop the carriage and the impulsive Losberne bursts his way in. They encounter 'a little, ugly hump-backed man', who is understandably furious at the invasion of his property by these two strangers and vociferates horrific but comically impotent threats. There is, as it turns out, absolutely nothing to prove that the burglars were ever in the house. The furniture and interior decoration are entirely different from what Oliver remembers and has told his friends about: 'not even the position of the cupboards; answered Oliver's description!' (p. 250). Oliver and the doctor leave with the little man's curses ringing in their ears. Losberne is embarrassed by the whole episode, and evidently sees it as evidence of Oliver's extraordinary nervousness. None the less, the good-hearted doctor trusts the boy sufficiently to go looking for Brownlow's house, which they discover after a little trial and error. Oliver remembers the way to the street and recognizes the house immediately by its white colour.

As Williamson points out, the odd thing about the business with the little ugly hump-backed man's house 'is its apparent pointlessness' (p. 226). But, Williamson suggests,

> if we approach *Oliver Twist* as a crime thriller, the obvious explanation of the confusion over the house is that the hunchback is an associate of Sikes who allows him the use of the house for his nefarious purposes, and that Dickens had planned that Losberne's action in entering the house should give its occupant a chance to see and identify Oliver . . . All the members of the gang would doubtless have been alerted to Oliver's disappearance in the district and the threat he constituted to their safety; and it would be easy enough for an astute hunchback to track Oliver down through his companion and the carriage he occupied. (pp. 227–8)

It is an attractive hypothesis—except that a bona fide thriller-writer would surely have alerted the reader to the significance of the event later in the story. Williamson implies that pressure of serialization may have prevented Dickens from working out this detail of the plot satisfactorily.

There are other logical objections. It would hardly be necessary for the gang to use accidental sources of information and all the complicated business of trailing carriages many miles into the countryside—assuming that the little man could set the operation up before Losberne's carriage was on its way into the maelstrom of the London streets. Since the wounded Oliver was taken into the same house that was set up for the burglary (an establishment that had been thoroughly 'cased' in advance), and had remained there several weeks convalescing from his bullet-wound, he could have been effortlessly tracked by Sikes, who could have found his way to the scene of the crime blindfolded. It is true that the Maylies have moved for the summer to the country; but they have left servants in the house who know the other address and have no reason for keeping it secret. It would be the work of minutes for the Artful Dodger to invent some ruse for being told where the Maylie household and their little invalid guest are now residing. If there is a larger significance in the episode of the little hump-backed man it is, surely, that despite this evidence of Oliver's unreliability (and his whole story strains credulity to breaking-point) the good doctor and his friends persist in believing him. Although he clearly is in error about the gang's house, they trust that he can locate Brownlow's house. And, by implication, they believe his whole incredible story.

Yet another explanation of the window episode is offered by Fred Kaplan in *Dickens and Mesmerism* (Princeton, 1975). Kaplan notes that *Oliver Twist* was written at the height of 'The Mesmeric Mania', when Dickens was closely associated with the arch apostle of this new science, Dr John Elliotson. The long disquisition about Oliver's half-sleeping sensory awareness seems a clear pointer to the author's current fascination with mesmerism and 'animal magnetism'. As Kaplan records, it was one of Elliotson's claims that 'the mesmerized subject can see with his eyes closed' (p. 152). This would seem to be how Oliver becomes aware of the criminals at the window before he has woken from his sleep—which may more truly be described as a trance, or what the mesmerists called 'sleepwaking'. One could go one step further (as Kaplan does not) and suggest that the whole episode is a mesmeric phenomenon. This would explain the lack of any footprints or visible signs of Fagin's and Monks's preternaturally sudden disappearance ('It was but an instant, a glance, a flash, before his eyes; and they were gone').

There is, as it happens, strong supporting evidence for the hypothesis that the whole episode is an example of what the practitioners of mesmerism called 'mental travelling'. It is not the case that Monks and Fagin visit Oliver; he visits *them*, borne on the wings of mesmeric trance. As Alison Winter has recorded, from August 1837 to May 1838 Elliotson carried out private experiments on many of his patients, and in particular the domestic servant, Elizabeth O'Key, in the wards of University College Hospital. In the spring and summer of 1838 he put on a series of public demonstrations at UCH. According to Kaplan, Dickens attended the first O'Key demonstration on 10 May 1838, or the second on 2 June, or 'perhaps even both' (p. 36). As Winter describes the experiments: 'Elliotson did things such as mesmerizing her through walls from various distances; she had visions in which she represented herself as if she felt that various personages were around her—these individuals told her things which became personal prophecies.'[5] There is also a notable similarity between the language of Dickens's remarks about Oliver's tranced sensitivity to absent personages, and what commentators were saying about O'Key in 1838.

Probably no explanation of this episode will convince everyone and some will convince no one. I would like, however; to offer an explanation of my own. *Oliver Twist* was first published as a serial in *Bentley's Miscellany*, from February 1837 to April 1839. It was an amazingly busy period in Dickens's early career. He had outstanding contracts for new novels and editorial commitments to no less than three different publishers, and felt that he was in danger of 'busting the boiler'.

One of the problems for the serialist working at full stretch was providing early enough copy for his illustrator, particularly if his partner (like George Cruikshank) needed to have his designs engraved on steel—a long and difficult procedure. When he had time in hand, Dickens preferred to supply manuscript or proofs to Cruikshank, so that he could portray narrative details accurately. But, as Kathleen Tillotson notes, 'although Dickens originally promised to let Cruikshank have the manuscript by the fifth of the month, the evidence suggests that after the first month he was always later, sometimes sending an instalment of manuscript, and sometimes conveying instructions for the illustration by a note or by word of mouth'.[6]

'Monks and the Jew' appeared in the *Bentley's* instalment for June 1838. Dickens felt himself particularly pressured at this point, because he thought copy was needed early, on account of the Coronation on the 28th of the month. My speculation is that before actually writing this section of the narrative Dickens foresaw an abduction or murder attempt on Oliver; and duly instructed Cruikshank to go ahead with the villains-at-the-window illustration, preparatory to that scenario. But, while writing the episode, Dickens settled on something more complex, bringing the Bumbles, Noah Claypole, and Bill Sikes back into the centre of things. It remains uncertain, *if* they are real, and not figments of Oliver's superheated imagination, what Fagin and Monks intend to do with the intelligence that Oliver is now lodged with the Maylies. But once Cruikshank had supplied the illustration, it was impossible, at this short notice, to procure another and Dickens suddenly realized that he could elegantly write himself out of the dilemma by means of the 'mesmeric enigma' device, using material gathered at the O'Key demonstrations.[7]

It would seem that Dickens's more scientific contemporaries registered the interesting overtones of the window scene in Chapter 34. G. H. Lewes wrote a letter (which has not survived) evidently enquiring exactly what Dickens had intended, and on what scientific authority the scene was devised. Towards the end of November 1838 Elliotson himself responded. Dickens wrote the following note to his illustrator:

My Dear Cruikshank,

Elliotson has written to me to go and see some experiments on Okey [*sic*] at his house at 3 o'clock tomorrow afternoon. He begs me to invite you. Will you come? Let me know.

Ever Faithfully Yrs.[8]

Why, one may go on to wonder, did Dickens in later prefaces to the novel not alert the reader to the scientific plausibility of Oliver's clairvoyance as, for instance, he ferociously defended the 'spontaneous combustion' in *Bleak House*, when G. H. Lewes questioned it? There is a likely explanation. In September 1838 O'Key was denounced as an impostor by Thomas Wakley in the *Lancet*. In the squabble that followed, Elliotson was forced to resign his position at UCH in late December of the same year. In the judgement of most intelligent lay persons (even those like Dickens who were friendly with Elliotson) the O'Key experiments, if not wholly discredited, had been rendered extremely dubious. In these circumstances, although he saw no reason to change his text, neither did Dickens see any good reason for drawing the reader's attention to the 'science' on which the window scene is based.

The Oxford World's Classics *Oliver Twist* is edited by Kathleen Tillotson with an introduction by Stephen Gill.

Charles Dickens · *Martin Chuzzlewit*

Mysteries of the Dickensian year

Novelists like George Eliot, Thackeray, and Trollope make it easy for the pedantic reader to work out a monthly calendar of events in their major fictions. In *Middlemarch*, for example, we can set down the following chronology for the novel's main events:

1829 early summer: Dorothea and Casaubon meet
1829 September: Dorothea and Casaubon marry
1829 Christmas: Dorothea and Casaubon in Rome
1830 May: Peter Featherstone dies
1830 July–August: Lydgate and Rosamond marry
1831 March: Casaubon dies
1832 March: Raffles dies
1832 May: Will and Dorothea marry
1832 June: the Reform Bill[1]

Vanity Fair, although it covers a much longer tract of history (1813–32) than *Middlemarch* (1828–32), is meticulous about calendar time-markings, particularly in its early, tight-knit sections covering the period from summer 1813 to summer 1815. Thackeray's narrative opens with ostentatious chronological exactitude: 'While the present century was in its teens, and on one sunshiny morning in June . . .' In the next hundred pages we learn that Rebecca is to stay ten days with the Sedleys, that the year coyly given as the 'teens' of the century must be 1813, and the 'sunshiny morning in June' is the fifteenth of that month.[2] Thackeray's chronometer continues its exact calendric beat throughout the subsequent narrative. As did George Eliot for *Middlemarch*, Trollope drew up a detailed monthly time-line for *The Way We Live Now.*[3] George Eliot made fine changes to her schedule. And it has been argued that Trollope used an actual 1872 calendar to get even greater precision for his layout of *The Way We Live Now.*[4]

Dickens is much less precise about monthly and seasonal references, particularly in his early novels.[5] Anyone attempting to draw up a calendric schedule for the central events of *Martin Chuzzlewit* will run into some perplexing and thought-provoking anomalies. These anomalies witness less to any carelessness on Dickens's part, than to his Shakespearian confidence in making the elements do whatever it is that the current mood and dramatic needs of his narrative require them to do. Dickens is no slave to the calendar.

To demonstrate this, one may start with Martin's break with Pecksniff. We are told the hero leaves on 'a dark winter's morning' (p. 179). There is supporting evidence as to the wintriness in Dickens's description of the bleak skies, cold rain, and mud that accompanies Martin's tumbril-like journey by cart from Salisbury to London. In Chapter 14, preparatory to his departure for the New World, Martin has his farewell meeting with Mary Graham in St James's Park. Dickens gives another wintry picture of the dawn assignation between the lovers:

> He was up before day-break, and came upon the Park with the morning, which was clad in the least engaging of the three hundred and sixty-five dresses in the wardrobe of the year. It was raw, damp, dark, and dismal; the clouds were as muddy as the ground; and the short perspective of every street and avenue, was closed up by the mist as by a filthy curtain.
>
> 'Fine weather indeed', Martin bitterly soliloquized. (p. 202)

The theme of funereal winter darkness is reiterated in the description of England, as Martin and Mark set sail for America (Chapter 15):

A dark and dreary night; people nestling in their beds or circling late about the fire; Want, colder than Charity, shivering at the street corners; church-towers humming with the faint vibration of their own tongues, but newly resting from the ghostly preachment 'One!' The earth covered with a sable pall as for the burial of yesterday; the clumps of dark trees, its giant plumes of funeral feathers, waving sadly to and fro: all hushed, all noiseless, and in deep repose, save the swift clouds that skim across the moon, and the cautious wind, as, creeping after them upon the ground, it stops to listen, and goes rustling on, and stops again, and follows, like a savage on the trail. (p. 211)

Dickens does not give any precise monthly reference, but it would seem self-evident from the bitter cold that Martin's embarkation occurs at deep midwinter, January or early February at the latest.

At one early point in New York, Martin tells the obnoxious Jefferson Brick that it is 'five weeks' (p. 225) since he left England. He and Mark cool their heels (although there are no precise references to seasonal temperature) for a few weeks more in the big city before setting off for Eden. No date reference is given. Meanwhile, back in Britain, old Anthony Chuzzlewit in Chapter 18 complains at 'What a cold spring it is!' (p. 255). Presumably a couple of months have passed and we are now to understand that it is early April. The same spring season is subsequently described as 'lovely', as Mr Pecksniff and Jonas return from Anthony's funeral (p. 285). The day of that ceremony is 'fine and warm'—late April, presumably Chapter 20 (in which this description occurs) ends Number 8 in the novel's original serialization, and Number 9 switches to America where, at exactly the same moment, it is implied, Mark and Martin are setting out on their journey to the interior. When the young men arrive in Eden (i.e. Cairo, Illinois, at the junction of the Ohio and Mississippi rivers), it seems to be high summer: 'A fetid vapour, hot and sickening as the breath of an oven, rose up from the earth, and hung on everything around' (p. 328). Illinois does not get baking hot until June and July.

Back in England, in what is now evidently midsummer, Pecksniff makes his wooer's assault on Mary. Dickens expatiates on the seasonal warmth and fecundity which matches Pecksniff's own bounding libido:

> The summer weather in [Pecksniff's] bosom was reflected in the breast of Nature. Through deep green vistas where the boughs arched over-head, and showed the sunlight flashing in the beautiful perspective; through dewy fern from which the startled hares leaped up, and fled at his approach; by mantled pools, and fallen trees, and down in hollow places, rustling among last year's leaves whose scent was Memory; the placid Pecksniff strolled. By meadow gates and hedges fragrant with wild roses . . . (p. 413)

Pecksniff's subsequent failure to win Mary's heart leads to Tom Pinch's dismissal, the crisis coming 'one sultry afternoon, about a week after Miss Charity's departure for London' (p. 419). Banished forever from his patron's favour, Tom also makes his way to London, by stage-coach. The panorama of pre-railway rural England that accompanies his trip lingers lovingly on the summery landscape which Tom sees from his seat on the driver's box:

> Yoho, among the gathering shades; making of no account the deep reflections of the trees, but scampering on through light and darkness, all the same, as if the light of London fifty miles away, were quite enough to travel by, and some to spare. Yoho, beside the village-green, where cricket-players linger yet, and every little indentation made in the fresh grass by bat or wicket, ball or player's foot, sheds out its perfume on the night. (p. 482)

Meanwhile, Martin and Mark undergo their regeneration in Eden. No precise time references are given, other than that after some weeks Martin falls ill, and it is thereafter 'many weeks' before he recovers from his 'long and lingering illness' (p. 455) sufficiently 'to move about with the help of a stick and Mark's arm' (p. 451). After they have resolved to leave Eden it is a further 'three crawling weeks' before the steamboat arrives to pick up the reluctant immigrants. On the voyage back along the river, it is clearly high summer: the first person Martin and Mark see on board the paddle-boat is 'a faint gentleman sitting on a low camp-stool . . . under the shade of a large green cotton umbrella' (p. 457). There is little delay, apparently, in their embarkation from New York, on the *Screw* again. The description of their arrival back in England in Chapter 35 is euphoric, with weather to match. They left at gloomy midnight; they return at joyous midday:

> It was mid-day, and high water in the English port for which the Screw was bound, when, borne in gallantly upon the fulness of the tide, she let go her anchor in the river.
>
> Bright as the scene was; fresh and full of motion; airy, free, and sparkling; it was nothing to the life and exultation in the breasts of the two travellers, at sight of the old churches, roofs and darkened chimney-stacks of Home. (p. 471)

The narrative goes on to inform us: 'A year had passed, since those same spires and roofs had faded from their eyes.' We feel, of course, that a year *must* have passed, for all this travelling, new experience, chronic illness, and moral regeneration to have taken place (apart from anything else, Martin on his return is clearly recovered in health). Adding up all the casual references to 'months', 'weeks', and 'days', a year might even seem too short. Primitive calculation, however, reveals that this twelve-month duration means that Martin and Mark must return in winter, not summer. Nor, on the other hand, can it be eighteen months (which might bring them to the next summer), since that would create a missing year in the convergent Pinch—Chuzzlewit sector of the narrative.

From the lack of any clear date-markers, we are given to understand that the travellers' return from the New World is coincidental with Tom Pinch's arrival in London, his reunion with Westlock and Ruth, and his installation with Mr Fips. Tom is well established in his new employment at the Temple while it is still summer. In Chapter 40, for instance, he is described going down to the London docks at seven in the morning, with Ruth, to enjoy 'the summer air' (p. 532).

The fact that these two strands of the narrative (the Martin strand and the Tom strand) have merged in summer (of whatever year it may be) is clinched by the great summer storm of Chapter 42. The prelude to this cataclysm, which will bring the melodrama to the boil, is described as coming at the climax of a heatwave:

> It was one of those hot, silent nights, when people sit at windows, listening for the thunder which they know will shortly break; when they recall dismal tales of hurricanes and earthquakes; and of lonely travellers on open plains, and lonely ships at sea struck by lightning. Lightning flashed and quivered on the black horizon even now; and hollow murmurings were in the wind, as though it had been blowing where the thunder rolled, and still was charged with its exhausted echoes. But the storm, though gathering swiftly, had not yet come up; and the prevailing stillness was the more solemn, from the dull intelligence that seemed to hover in the air, of noise and conflict afar off. (p. 550)

On the night of the storm Mark and Martin come to the Blue Dragon. This picks up a comment made by Mark immediately on returning, as they sit in the tavern by the docks: 'My opinion is, sir . . . that what we've got to do, is to travel straight to the Dragon' (p. 472).

This, then is the problem: some six hectic months have passed in Tom Pinch's life, bringing him from February to August. 'A year' has passed in Mark and Martin's life bringing them from February to the same August. Clearly, if Dickens had been precise about his chronology, Martin and Mark would have missed the great summer storm by six months, arriving back in gloomy, wintry January or February following. Nor could their stay in America be abbreviated to half-a-year, given all the events and long experiences it had to contain. 'There are some happy creeturs', Mrs Gamp observes to Mr Mould, 'as time runs back'ards with' (p. 347). Martin Chuzzlewit would seem to be just such a happy creetur.

Dickens eludes any charge of error by a kind of prophylactic vagueness—he never names a month, only seasons. But the missing six months in his hero's career also relates to the Great Inimitable's masterful way with background.[6] Clearly, for the purposes of mood, Dickens wanted black, depressive winter for Martin's departure for the New World. It matched the low point his career had reached. As clearly, Dickens wanted high summer for his hero's return. In its turn, the summer season chimed with Martin's spiritual rebirth and his happier relationship with life, society, and his friends and sweetheart. Coincidentally, one may note that Martin's trip to America matches Dickens's exactly: the novelist left Liverpool on 2 January 1842, and returned in June the same year.

The Oxford World's Classics *Martin Chuzzlewit* is edited by Margaret Cardwell.

Emily Brontë · *Wuthering Heights*

Is Heathcliff a murderer?

When he returns to Wuthering Heights after his mysterious three-year period of exile Heathcliff has become someone very cruel. He left an uncouth but essentially humane stable-lad. He returns a gentleman psychopath. His subsequent brutalities are graphically recorded. They are many and very unpleasant. He humiliates Edgar Linton who has married Cathy during his absence. 'I wish you joy of the milk-blooded coward' (p. 115), he tells Cathy in her husband's presence. The taunt is the more brutal since Edgar is clearly the weaker man and in no position to exact physical reparation. Heathcliff goes on to torment Edgar by hinting that he has cuckolded him. Subsequently Heathcliff beats his wife Isabella, as he has gruesomely promised to do in earlier conversation with Cathy: 'You'd hear of odd things, if I lived alone with that mawkish, waxen face; the most ordinary would be painting on its white the colours of the rainbow, and turning the blue eyes black, every day or two; they detestably resemble Linton's' (p. 106).

When Nelly sees Isabella, after she has fled from Heathcliff, she does indeed describe 'a white face scratched and bruised' (p. 170). Isabella goes on to describe her husband's 'murderous violence' (p. 172) to Nelly in some detail. Heathcliff has shaken her till her teeth rattle (p. 170). He has thrown a kitchen knife at her head which 'struck beneath my ear'; she has a wound which will probably scar her for life (p. 181). Had she not run away, who knows how far he would have gone in his cold brutality towards her.

In later life Heathcliff would certainly have beaten his son as savagely as he beat the boy's mother, were it not that he needs the degenerate brat whole and unmarked for his long-term scheme of revenge against Thrushcross Grange. He has no compunction about punching young Catherine. Young Heathcliff tells Nelly about his father's violent reaction on learning that the girl has tried to keep for herself two miniatures of her dead parents:

> 'I said *they* were mine, too; and tried to get them from her. The spiteful thing wouldn't let me; she pushed me off, and hurt me. I shrieked out—that frightens her—she heard papa coming, and she broke the hinges, and divided the case and gave me her mother's portrait; the other she attempted to hide; but papa asked what was the matter and I explained it. He took the one I had away, and ordered her to resign hers to me; she refused, and he—he struck her down, and wrenched it off the chain, and crushed it with his foot.'
>
> 'And were you pleased to see her struck?' I asked: having my designs in encouraging his talk.
>
> 'I winked,' he answered. 'I wink to see my father strike a dog, or a horse, he does it so hard.' (p. 281)

Or a woman, one may add. It is not just four-footed victims who feel the weight of Heathcliff's fist.

Heathcliff is capable of more cold-blooded and calculating cruelty. He abducts young Catherine and keeps her from her dying father's bedside, accelerating Edgar's death and ensuring that it shall be an extremely miserable one. He urges Hindley towards self-destruction by encouraging his fatal mania for drink and cards. On a casual level, Heathcliff is given to killing household pets (he strangles his wife's favourite dog by way of wedding present) and desecrates graves.

Mr Heathcliff; we may assume, is not a nice man. And in a later age his violence and lawlessness would have earned him a prison sentence—or at the very least a string of restraining orders and court injunctions. But does Heathcliff commit the cruellest crime of all, murder?

To answer this question we must examine the suspicious circumstances of the death of Hindley Earnshaw, master of Wuthering Heights. 'The end of Earnshaw was what might have been expected,' Nelly recalls in her long narrative to Lockwood, 'it followed fast on his sister's, there was scarcely six months between them. We, at the Grange, never got a very succinct account of his state preceding it.' Nelly learns of the death, after the event, from the local apothecary, Mr Kenneth. 'He died true to his character;' Kenneth cheerfully adds, 'drunk as a lord' (p. 184). Hindley was just 27. Evidently Kenneth has witnessed the death and signed the necessary certificate.

Nelly's suspicions are immediately aroused. 'Had he fair play?' she ponders. The anxiety 'bothers' her and she makes a trip to Wuthering Heights to discover what she can of the truth of the case. Before going she learns from Earnshaw's lawyer (who also acts for Mr Linton, Nelly's employer) that the 'whole property [of Wuthering Heights] is mortgaged'—to Heathcliff.[1] At the Heights, Nelly meets Heathcliff who, rather shiftily, as we may think, gives his eyewitness account of Hindley's death:

> 'That fool's body should be buried at the cross-roads, without ceremony of any kind [i.e. Hindley committed suicide]—I happened to leave him ten minutes, yesterday afternoon; and, in that interval, he fastened the two doors of the house against me, and he has spent the night in drinking himself to death deliberately! We broke in this morning, for we heard him snorting like a horse; and there he was, laid over a settle—flaying and scalping would not have wakened him—I sent for Kenneth, and he came; but not till the beast had changed into carrion—he was both dead and cold, and stark; and so you'll allow, it was useless making more stir about him!' (p. 185)

By the last enigmatic remark, Heathcliff means that it

would have been 'useless' calling in the coroner, on the grounds that the death was suspicious.

Heathcliff's account is 'confirmed' to Nelly by Joseph, the misanthropic (but wholly reliable) old manservant at the Heights. Joseph, however, is by no means happy about his former master's last hours:

> Aw'd rayther he'd goan hisseln fur t'doctor! Aw sud uh taen tent uh t'maister better nur him—un' he warn't deead when Aw left, nowt uh t'soart' ['I would rather that Heathcliff had gone himself for the doctor! I should have taken care of the master better than him—and he wasn't dead when I left, nothing of the sort!'] (pp. 185–6).

Joseph is invincibly honest. And one concurs in his 'muttered' doubts (he dare not voice them out loud, in case Heathcliff hears, and gives him the back of his hand). It is most improbable that a 27-year-old man, in otherwise robust health, should be able to 'drink himself to death' in a single night. Young men do, of course, kill themselves by excessive drinking, but usually by driving cars drunk, or by inhaling their own vomit while sleeping. It is clear that—although he is 'snorting'—Hindley is breathing efficiently when he is left alone with Heathcliff. Did he show signs of being about to suffocate, it would be an easy thing for Heathcliff to lift him up and bang him on the back, thus clearing his throat. And, as Joseph recalls, although dead drunk, Hindley did not appear to be dying. He was, however, insensible and incapable of resisting anyone stifling him with a cushion. Kenneth is a somewhat elusive figure, but it is likely that as a mere apothecary ('Mr' Kenneth) he would not have been able to conduct any expert medical examination of the body. It may even be that Heathcliff bribed him to sign the certificate and obviate any embarrassing coroner's inquest.

It is nicely poised and every reader must make his or her own judgement. If Heathcliff did stifle Hindley (albeit that Hindley has earlier tried to shoot and stab Heathcliff) we have to see him as a sociopathic monster. If he watched the man die, and declined to prevent his death (by clearing Hindley's throat, for example) he is scarcely better. These plausible reconstructions of what happened at Wuthering Heights while Heathcliff and the incapable Hindley were alone together render absurd such rosy adaptations as the Samuel Goldwyn 1939 film (the Goldwyn screenplay, by Ben Hecht and Charles MacArthur, ends with Heathcliff, played by Laurence Olivier, and Cathy, played by Merle Oberon, reunited as carefree ghosts skipping merrily over Penistone Crags). If we believe that Heathcliff was simply an innocent bystander at Hindley's self-destruction, then we can credit the sympathetic reading of his character suggested by the exclamation Nelly overhears him make, in the intensity of his wretchedness: 'I have no pity! I have no pity! The [more the] worms writhe, the more I yearn to crush out their entrails! It is a moral teething, and I grind with greater energy, in proportion to the increase of pain' (p. 152).

When a baby savagely bites its teething ring, it is because it (the baby) is experiencing excruciating pain from the teeth tearing their way through its gums. So Heathcliff may be seen to inflict pain on others (hurling knives at his wife, taunting Edgar, striking young Catherine, lashing his horse) only because he feels greater inward pain himself. But one cannot so justify the furtive smothering, in cold blood, of someone whose death will mean considerable financial gain to the murderer.

There are no clear answers to this puzzle. As Ian Jack has noted, '*Wuthering Heights* is one of the most enigmatic of English novels'. Whether or not Heathcliff is guilty of capital crime remains a fascinating but ultimately inscrutable enigma at the very heart of the narrative. For what it is worth, I believe he *did* kill Hindley, although for any unprejudiced jury it is likely that enough 'reasonable doubt' would remain to acquit him.

The Oxford World's Classics *Wuthering Heights* is edited by Ian Jack with an introduction by Patsy Stoneman.

Charlotte Brontë · *Jane Eyre*

Rochester's celestial telegram

On the face of it, Rochester's astral communication with the heroine at the conclusion of Jane Eyre ('Jane! Jane! Jane!', p. 419) is the most un-Brontëan thing in Charlotte Brontë's mature fiction. This was the author who declared in the preface to one of her novels that it should be as unromantic as a Monday morning. The Jane—Rochester exchange across the ether would seem to be the stuff of *Walpurgisnacht.* It is the more surprising since Brontë is a novelist who firmly eschews supernatural agency and intervention in her narratives.

It will help to summarize the events which precede Rochester's celestial telegram. Jane is at St John Rivers's home, where she is detained after evening prayers. It is around nine o'clock on a Monday evening. The family and servants go to bed. St John renews his proposal of marriage. Jane wavers: 'I could decide if I were but certain,' she says, 'were I but convinced that it is God's will I should marry you' (p. 419). She wants a sign. There are only a few minutes-worth of conversation recorded, but evidently St John Rivers and Jane are together for a period of some hours, she staring intently at the 'one candle' which illuminates the room (the Rivers' household is frugal):

> All the house was still; for I believe all, except St John and myself, were now retired to rest. The one candle was dying out: the room was full of moonlight. My heart beat fast and thick: I heard its throb. Suddenly it stood still to an inexpressible feeling that thrilled it through, and passed at once to my head and

extremities. The feeling was *not like an electric shock*; but it was quite as sharp, as strange, as startling: it acted on my senses as if their utmost activity hitherto had been but torpor; from which they were now summoned, and forced to wake. They rose expectant: eye and ear waited, while the flesh quivered on my bones.

'What have you heard? What do you see?' asked St John. I saw nothing: but I heard a voice somewhere cry—

'Jane! Jane! Jane!' Nothing more. (p. 419, my italics)

Jane recognizes the 'known, loved, well-remembered voice' of Edward Fairfax Rochester. She rushes into the garden and calls back, 'I am coming! Wait for me! Oh, I will come!'

Brontë moves quickly to forestall any 'Gothic' interpretations by her readers. 'Down superstition!', Jane is made to command: 'This is not thy deception, nor thy witchcraft: it is the work of nature. She was roused, and did—no miracle—but her best' (p. 420). That Rochester's communication was not hallucinatory is confirmed after Jane makes her trip to Ferndean four days later. Rochester tells her that on the Monday night in question he sat for some hours in his room gazing at the moon (as Jane was simultaneously gazing at the candle). Involuntarily, 'near midnight' he came out with the exclamation 'Jane! Jane! Jane!' Then, to his consternation, he heard her voice reply, 'I am coming: wait for me!' (p. 447). He does not, at this point, know that she heard his call and used in reply the very words which at his end he reports hearing.

What is going on here? Margaret Smith's note to the previous World's Classics edition puts the problem clearly, if inconclusively:

Charlotte Brontë defended this incident by saying 'But it is a true thing; it really happened.' (See *Life*, ii. 149) A similar incident occurs in the Angrian story of*Albion and Marina* (1830) as Miss Ratchford points out: see *B.S.T*, 1920, p. 13 f., and Ratchford 212. Charlotte Brontë's fear of being accused of plagiarism when, after she had written *Jane Eyre*, she read of the midnight voice in Mrs Marsh's *Two Old Men's Tales*, certainly rules out any conscious literary indebtedness. See *Life*. ii. 311. Parallels have been noted in Defoe's *Moll Flanders* and George Sand's *Jacques* (by Mrs Humphry Ward, Haworth edition, 1899).

It is not hard to come up with other literary analogies, some of which Brontë might have been more likely to know than Defoe (the voice which St Augustine in his *Confessions* recalls hearing; the visionary communication between the tragic lovers in Keats's *Isabella*). But what is problematic is the author's insistence, reiterated in the text and in commentary on the text, that 'it is a true thing; it really happened'; 'no miracle.' Brontë does not, as would seem tempting for someone perpetrating an episode so fantastic, take refuge in the traditional licence of the romancer.

Some early reviewers apprehended that Brontë was alluding to the newly discovered invention of telegraphy.[1] But 1847 is too early for this. It is true that the Electric Telegraph Company was formed in 1846, but it was not until the 1850s that full commercial exploitation occurred. What seems more likely is that Brontë was drawing on her knowledge of the science of mesmerism. She was, as is well recorded by her biographers, fascinated by 'animal magnetism' (as mesmerism was sometimes termed in the 1840s) and the related field of phrenology. She attended lectures on the subject in the early 1840s and communicated with mesmeric investigators and phrenologists. Her novels are peppered with incidental allusions to her knowledge of the field.

In her defiant assertion, 'it is a true thing', Charlotte Brontë was probably thinking of two specific authorities on the subject of mesmerism and clairvoyant communication. Catherine Crowe, a novelist and popularizer of spiritualism, had recently attempted a historico-scientific vindication of psychic phenomena in her translation of the German *Die Seherin von Prevorst* (*The Seeress of Prevorst*, translated from the German of Justinus Kerner, London, 1845). The 'seeress' was Frederica Hauffe, born in 1801, whose life was a long succession of witnessed and confirmed acts of clairvoyance, prevoyance, 'sleep-seeing', and prophecy—all justified by Crowe on the basis of 'magnetism' (the explanatory source of mesmerism). In 1848 Crowe published *The Night Side of Nature*, an assortment of weird tales, haunted houses, supernatural happenings, and apparitions—all vouched for as genuine by the author.[2]

Crowe, who was to become a full-time propagandist for spiritualism in the 1850s (and a social acquaintance of Brontë's, after the success of *Jane Eyre*), was probably congenial to Brontë but less than convincing as a scientific authority. In making her claim for the 'truth' of the etherial exchange between Jane and Rochester, Brontë was more likely drawing on another popular treatise on mesmeric phenomena, the Revd Chauncy Hare Townshend's *Facts in Mesmerism, with Reasons for a dispassionate Inquiry into it* (London, 1840). Townshend's book, which was put out by the very respectable publisher Longman and Green, offered a huge array of 'facts' testifying to the validity of the pseudo-science. Many of these facts dealt with remote communication between mesmerically sensitized subjects. Townshend was quite dogmatic that this was a scientifically authenticated truth and cited chapter-and-verse cases to support his theories:

> It has been said that persons in certain states either mesmeric or akin to the mesmeric can become aware of the thoughts of others without the usual communication of speech . . . But is there, it may still be asked, any one acknowledged instance in nature by which the possibility of receiving actual experiences, other than

by the normal inlets of sense can be demonstrated? There is. (pp. 365, 460)

Townshend gives a number of historically recorded instances of messages received 'other than by the normal inlets of sense'. The 'Testimony to a curious Fact by Dr Filippi of Milan' in July 1839, for example: 'Mr Valdrighi, advocate, had his sense of hearing so exquisitely exalted that he could hear words pronounced at the distance of two rooms, the doors of which were shut, although pronounced in a weak and low voice' (p. 473).

More significantly, Townshend, who was himself a practising mesmerist, describes experiments along these lines which he conducted with a subject called 'Anna M.' He discovered that it was possible to 'magnetize' her from a distance—and communicate with her, suggesting objects and messages which she could pick up far beyond the range of the human ear. Over the course of his experiments with Anna M., Townshend successfully extended the distance between himself and his magnetized subject to a quarter of a mile, transmitting his 'influence' electrically (as he assumed) and communicating information to her,

There were other sensational displays of mesmeric clairvoyance and 'mental travelling' which might have inspired Brontë's protestation that the 'Jane! Jane! Jane!' episode was a 'true thing'. In the early 1840s Alexis Didier (previously a clerk in a Parisian haulage firm) toured extensively in France, Belgium, and England, giving shows. As Alison Winter records: 'His routine included playing cards and reading books whilst blindfolded, identifying the contents of envelopes, and "travelling clairvoyance"— viewing objects at a great distance.'[3] Even more likely to have been in Brontë's mind was the public quarrel over clairvoyance and mental travelling that Harriet Martineau became involved with in late 1844 and early

1845. The Martineaus (the woman of letters, Harriet, and her Unitarian brother, James) were the most powerful literary figures with whom the young Brontës could claim connection before the success of *Jane Eyre* made Charlotte nationally famous.[4] In late November 1844 Harriet Martineau sent the London review; the *Athenaeum*, a long letter describing the clairvoyant feats of a young maid-servant called Jane Arrowsmith (the echo of this name in 'Jane Eyre' may not be accidental). Sensationally, Martineau claimed that her Jane had, while in a mesmeric trance, witnessed a shipwreck occurring some dozen miles distant at sea. Jane also gave proof of being able to hear at great distances, without the aid of her physical ears. Martineau followed up with three more pieces, describing and analysing Jane's powers of 'mental travelling' as a demonstration of the truth of mesmeric science.[5]

These reports of Jane Arrowsmith's clairvoyance provoked a huge controversy, including pamphlets and a series of savagely denunciatory pieces by an irate London doctor, who visited the town where Jane Arrowsmith lived and claimed to have discovered that she had been told about the shipwreck just before the crucial session with the mesmerist. Martineau stuck to her guns, insisting on the 'truth' of Jane's clairvoyance. It is extremely likely that, one way or another, Charlotte would have caught wind of the Jane Arrowsmith affair; it is equally likely that she would have stood firmly with Harriet Martineau in asserting its 'truth'.

How then should we read the critical scene in *Jane Eyre*? Accidentally, it would seem, both Jane and Rochester put themselves simultaneously into a mesmeric state of 'sleepwaking'. Jane does it by staring for a long period at the single candle (this, incidentally, was the standard parlour-game technique for putting someone under the influence; it is possible that Brontë had used it successfully on herself). Rochester produces the same effect on himself by staring at the moon. He, as the stronger will, has Jane under his influence, which she feels as something akin to, but not quite like, an electric shock (one of Townshend's favourite tricks at lectures on mesmerism was to give members of his audience mild electrical shocks, to demonstrate the nature of mesmeric energy). Like Anna M. and Townshend, in this condition of nervous 'exaltation' the exchange of messages can take place. It is, as Jane protests, no 'miracle', but an accident produced by their fortuitously mesmerizing themselves at the same critical moment.

The Oxford World's Classics *Jane Eyre* is edited by Margaret Smith with an introduction by Sally Shuttleworth.

W. M. Thackeray · *Vanity Fair*

Does Becky kill Jos?

Students of literature are routinely told that Thackeray is an 'omniscient' novelist; indeed, that with Fielding he is probably the perfect specimen of the type. He himself tells us, repeatedly and with apparently complacency in *Vanity Fair*, that 'the novelist knows everything'. But this omniscience has its holes. The reader is teased by what this allegedly all-knowing narrator would seem not to know, will not acquaint himself with, or declines to impart. Omniscient he may be; omnidictive he is not.

Most provoking of the text's silences is that concerning Jos's death. He dies in mysterious circumstances on the continent, sometime in the early 1830s, while in the dangerous company of Becky Crawley. From his first encounter with her, some twenty years before, Jos has been in danger from this fatal woman. In 1813 she almost netted him; but George Osborne, unwilling to have a governess marry into the family ('low enough already, without *her*', p. 71) frightened the fat man off. At Pumpernickel, despite Dobbin's efforts, she finally lands her prey. Becky cannot marry Jos (Rawdon, her estranged husband, is still staving off the fevers of Coventry Island). But she lives with her victim until he dies—prematurely. She is his insurance beneficiary; the rest of the nabob's once substantial wealth has mysteriously evaporated. And in later life Becky is a *very* prosperous lady, we are told. When she was first setting her hat at Jos in Russell Square she was netting a purse; now, at last, it would seem that the purse is comfortably full.

How does Jos die? The insurance people are suspicious. Their solicitor swears it is 'the blackest case that ever had come before him' (p. 877). Thackerayan innuendo confirms our sense that Becky helped Jos out of the world. *Her* solicitors are ominously named Messrs. Burke, Thurtell and Hayes. Burke, with Hare, was the Edinburgh body-snatcher who killed and sold corpses to the university school of medicine. John Thurtell was a murderer, hanged in 1824. Catherine Hayes was a husband killer, celebrated by Thackeray in his anti-Newgate satire, *Catherine* (1839).

There is another broad hint in the penultimate full-plate illustration to the novel, 'Becky's second appearance in the character of Clytemnestra' (p. 875).

Becky's second appearance in the character of Clytemnestra

Becky's first appearance as the Greek husband-killer was in the charade at Gaunt House, just before she betrayed Rawdon into the hands of the bailiffs. Here we feel that she will use the knife that, somewhat melodramatically, Thackeray shows her holding. (An ironic Hogarthian print of the good Samaritan is behind Jos, who vainly implores an implacable Dobbin to help him.)

It all points one way. But why does Thackeray not tell us straight out? It is a mote that he seems deliberately to have left to trouble generations of readers. And when asked in later life by just one such troubled reader; 'did Becky kill Jos?' the novelist is reported to have merely smiled and answered, 'I don't know'.

'Was she guilty?' The narrative asks that question of Becky (but gives no direct answer) at two crucial junctures: first, in the liaison with Steyne; secondly, after Jos's death. It is, of course, odds on that Becky was thoroughly guilty of both these and many other like offences. Would the notoriously lecherous Marquess of Steyne have given Becky a cheque for over a thousand pounds, provided for her son and companion Briggs, and given her diamonds (which she feels obliged to hide from her husband) if he were not enjoying with her what he more flagrantly enjoys with the Countess of Belladonna? So too with Jos's untimely decease; any open-minded reader concurs with the insurance office's suspicion.

If we accept the hint that Becky indeed killed Jos, then the last illustration 'Virtue rewarded: a booth in Vanity Fair' (with its ironic echo of Fielding's *Amelia, or Virtue Rewarded*) is one of the most un-Victorian endings in Victorian fiction. I can only think of one other Victorian novel in which a main character escapes punishment for murder (Mrs Archer Clive's eccentric romance, *Paul Ferroll*, 1855). To have left Becky unpunished for her capital offence would also have been radically out of character for Thackeray, who had been one of the main castigators of the so-called Newgate Novel—more particularly the 'arsenical' variety recently made notorious by Bulwer-Lytton's *Lucretia* (1846), a novel which *The Times* called 'a disgrace to the writer, a shame to us all'—on the grounds that it glorified wives who poisoned husbands for gain. Thackeray had built his early career around attacks on the immoralities of Bulwer-Lytton's fiction and its depictions of vice rewarded.

Virtue rewarded. A booth in Vanity Fair

Would Thackeray, one wonders, have emulated a writer whom he loathed? More significantly, as has been pointed out by a number of commentators on *Vanity Fair*, murder seems entirely out of character in Becky—an adventuress who might well stoop to some well-paid adultery but is, we feel, no homicidal psychopath capable of the premeditated crime of slow poisoning by arsenic.

To return to the text which surrounds the Clytemnestra illustration. In the last years of his life the 'infatuated man', Joseph Sedley, is reported to be entirely Becky's 'slave'. Colonel Dobbin's lawyers (who have clearly been undertaking some discreet spying on their client's behalf) inform him that Jos has taken out a heavy insurance upon his life. Moreover; 'his infirmities were daily increasing' (p. 873). What, one may well ask, are these 'infirmities'—the physical decrepitude consequent on a lifetime's gluttony? Or the slow effects of criminally administered toxins?

Dobbin, at his wife's alarmed request, goes to visit his brother-in-law in Brussels, where he is staying in an adjoining apartment to Mrs Crawley. She is living in great style, presumably on Jos's dwindling store of money. A mysteriously terrified Jos tells Dobbin that Mrs Crawley has 'tended him through a series of unheard-of illnesses, with a fidelity most admirable. She had been a daughter to him' (p. 873). He, despite these daughterly attentions, is perceived by the Colonel to be in 'a condition of pitiable infirmity'. Mrs Crawley, Jos further insists to a disbelieving Colonel, 'is as innocent as a child'. The Colonel leaves, sternly indicating that he and Mrs Dobbin can never visit such as Mrs Crawley and her consort again. Before doing so, he urges Jos 'to break off a connexion which might have the most fatal consequences to him' (p. 874). Three months later Jos duly dies at Aix-la-Chapelle, a watering place, whither he and Mrs Crawley have repaired in a vain attempt to recover his health.

There follows the coded business about the lawyers, Burke, Thurtell and Hayes, and the insurance company's dark suspicions.

If Becky killed Jos, how was it done? By poison? Or is he, as it seems in his last interview with Dobbin, terminally ill and terrified of dying alone? Someone, that is, who is going to depart the world without the assistance of arsenic. The Clytemnestra picture is, on close inspection, baffling. It is made clear in the text that Becky is not, in fact, hiding behind the screen. (Jos is morbidly careful to arrange the meeting 'when Mrs Crawley would be at a *soirée*, and when they could meet *alone*', p. 873.) Nor, if Becky actually does kill Jos, is the deed done with a knife. Whatever else, Becky is no Lizzie Borden.

What the picture would seem to allegorize are the exaggerated fears and suspicions of the respectable world ('I warrant the heartless slut was behind the screen all the time, just biding her time to kill the poor man!'). And Thackeray casts those suspicions in their most lurid form. So lurid, in fact, that the discriminating reader must dismiss them as preposterous. Of course—if we weigh up all the prior evidence Becky is no cutthroat. There is no question but that she is an unscrupulous woman, taking monetary advantage of a dying man, treating him doubtless with the same careless kindness which characterizes her last acts towards Amelia (whose path to a happy marriage she clears, with some well-placed malicious information about George Osborne).

In short, in this last section of the novel Thackeray is playing a game with his readers. He lures them—by flattering their responsiveness to authorial nods and winks—into thinking themselves cleverer than they in fact are. Complacently we readers, priding ourselves on being sophisticated enough to decode the Burke, Thurtell, Hayes, and Clytemnestra allusions, fall into the same vulgar prejudice as does the 'world' that condemns Becky. Does Becky kill Jos? Of course she doesn't—but maliciously wagging respectable tongues will never believe otherwise.

The Oxford World's Classics *Vanity Fair* is edited by John Sutherland.

Anne Brontë · *The Tenant of Wildfell Hall*

Who is Helen Graham?

Shortly after her arrival, the hero's irrepressibly cheeky young brother, Fergus, tells the fascinating new tenant of Wildfell Hall: 'It amazes me, Mrs Graham, how you could choose such a dilapidated, ricketty old place as this to live in. If you couldn't afford to occupy the whole house, and have it mended up, why couldn't you take a neat little cottage?' (p. 57). The lady gives an unsatisfactory answer—perhaps, she lightly tells the young man, she was too proud or too 'romantic'.

Fergus's observation opens up another question which neither Helen nor Anne Brontë's text answers for us. Helen Huntingdon wishes to escape her brutally alcoholic husband, Arthur. Given the date (and until 1857, well after the novel's publication), Mr Huntingdon would be quite entitled, should he find his errant wife in England, to repossess his conjugal rights (i.e. rape her at will), to repossess whatever personal things she has taken with her as his stolen property, and to reassume sole parental responsibility for his son, little Arthur. It is clear (since he has adamantly refused her permission to leave his house, has alienated their child's affections, and has removed her from all access to money) that Mr Huntingdon is in no mood to give up the chattels which the marriage laws of England have made over to him. Helen has every reason for going to ground and for throwing her vindictive spouse off her scent with every ingenious ruse available to her. To this end, she has changed her name to 'Mrs Graham', and has taken refuge in the broken-down mansion, Wildfell Hall. In her farewell letter to her closest friend, Esther Hargrave (whom she does not inform of her future whereabouts), Helen stresses that it is 'of the last importance that our future abode should be unknown to him [her husband] and his acquaintance'. She will disclose it to 'no one but my brother', she tells Esther (pp. 369–70). At one point, Helen seriously contemplates emigration to America to escape her husband's clutches.

With this in mind, it is very odd indeed that Helen chooses Wildfell Hall as her asylum. The house belongs to her brother, Frederick Lawrence, 'the young squire whose family had formerly occupied Wildfell Hall, but had deserted it, some fifteen years ago, for a more modern but commodious mansion in the neighbouring parish'. It would not take a moderately curious husband (or his lawyers) a week to discover a runaway wife and child (she masquerading as a widow, but going into local society with no other cover than an assumed name) in such an obvious hiding place as her parental home. It is clear that Mr Huntingdon does indeed institute a vigorous search for Helen and Arthur—why then does he not find her?

This goes together with other odd features in Helen's background. Although, as it emerges, her parents have just the two children, and clearly have lived as local dignitaries in Gilbert Markham's district of England (we are never precisely told where it is), it emerges that Helen has only seen her father once in her adult life and—until her flight from Grassdale—her brother is a virtual stranger to her. We learn this, obliquely, in one of the early marital quarrels with Arthur, when Helen dresses in black in recognition of her father's death, and indicates an intention to attend his funeral. 'I hate black', Arthur says:

> 'I hope, Helen, you won't think it your bounden duty to compose your face and manners into conformity with your funereal garb. Why should you sigh and groan, and I be made uncomfortable

> because an old gentleman in —shire, a perfect stranger to us both, has thought proper to drink himself to death?—There now I declare you're crying! Well, it must be affectation.'
>
> He would not hear of my attending the funeral, or going for a day or two, to cheer poor Frederick's solitude. It was quite unnecessary, he said, and I was unreasonable to wish it. What was my father to me? I had never seen him, but once since I was a baby, and I well knew he had never cared a stiver about me;—and my brother too, was little better than a stranger. (pp. 256–7)

That brother has, apparently, been brought up in the bosom of his family as the future squire. Helen, however, is quite unknown to anyone in the district. Although the Lawrence family inhabited Wildfell Hall until fifteen years ago (well within the memory of Gilbert's parents and the local vicar, the Revd Milward)—she has never, before reappearing as 'Mrs Graham', set foot in Wildfell Hall, nor, apparently, in the nearby commodious mansion.

There are some curious features here, which correspond with other curious features. Helen's mother is never mentioned in the text—she is simply not there (her funeral may be the 'one' occasion on which Helen has met her father in adult life). Helen's 'wedding to Arthur is only described summarily in passing and the ceremony, we apprehend, was not graced with the presence of either her father or brother. Although her family is demonstrably well off (they left Wildfell Hall not because of financial difficulties, but to take up a more luxurious residence), she has brought no dowry with her to the wedding, and only a tiny portion of family jewels (p. 337)—on the face of it, the only daughter of such a prosperous family, marrying with full parental consent, should come to her husband laden with treasure. Helen inherits nothing on her father's death, although her brother Frederick comes into considerable wealth. Helen was brought up by her uncle and aunt Maxwell (Mrs

Maxwell was her father's sister, apparently; see p. 456). It is they who bring her into society, vet her suitors, and make all necessary parental choices. When Mr Maxwell dies, he leaves—on his wife's instruction—the bulk of his fortune not to his widow (as, being childless, would he natural), but to Helen. We are told that Mrs Maxwell (as the former Miss Lawrence) brought a fine fortune—the bulk of her husband's subsequent wealth—to the marriage with her. It has apparently not always been the practice of the Lawrence family to send their daughters penniless to the altar.

It is singular that we are not informed by the text what Mrs Graham/Huntingdon's maiden name was. The narrative of Helen's premarital life (which is given through her own journal) goes to considerable contortions to avoid divulging this information in any clear way (see, for instance, Mr Boarham's laborious use of the 'My dear young lady' formulation, to avoid using the 'Miss Lawrence' address, which would be natural; pp. 132–4). The only way we can work out Helen's maiden surname is by speculation and the text (her text) gives us no firm evidence that she is, indeed, Miss Lawrence, the only daughter of Squire Lawrence (senior) the resident of Wildfell Hall, and the sister of Squire Lawrence (junior), the landlord of Wildfell Hall.

There are two hypotheses which the reader can advance to account for these anomalies. The first is that, the old Squire Lawrence being a dipsomaniac (see Huntingdon's comment about his drinking himself to death, and Frederick's enigmatic comments about congenital alcoholism on p. 38), his young daughter was removed as a baby from the house, to save her from its corrupting influence. But this would not explain her subsequent estrangement from her brother Frederick (who is clearly not a toper nor misanthropic), nor her apparent disinheritance from what would seem to be an only daughter's normal portion of family wealth.

The second hypothesis, which is more plausible, is that Helen is illegitimate—one of her debauched father's by-blows. As the central narrative makes clear, in the world of Brontë's novel a weakness for drink goes together with the grossest sexual delinquency. Helen, we surmise, is not Squire Lawrence's legal daughter but his bastard daughter, not Frederick's sister, but his half-sister. This would explain her total alienation from 'her' family and her being brought up well away from Wildfell Hall, and never once returning even for such family festivals as Christmas. It would also explain why Mrs Maxwell is at such pains to instill a high level of sexual morality into her ward. As a bastard, Helen would, in English law, take not her father's but her mother's surname which was, significantly, 'Graham'—the name by which she chooses to be known at Wildfell Hall. Illegitimacy (and the disowning that goes with it) would explain why her family does not attend her wedding, why her husband has met none of her family (other than the Maxwells), and why it is he cannot easily trace her back to Wildfell Hall.

The Oxford World's Classics *The Tenant of Wildfell Hall* is edited by Herbert Rosengarten, with an introduction by Margaret Smith.

Mrs Gaskell · *Mary Barton*

What kind of murderer is John Barton?

As Edgar Wright points out, Mrs Gaskell 'originally meant her novel to be called "John Barton". She mentions to two or three correspondents that she had envisaged the novel as a tragic poem with John Barton as the hero' (p. xv). This ur-narrative would have concentrated in detail on the working-class hero's suffering, his alienation and seduction by the 'vile' doctrines of Chartism and Communism, his lapse into homicidal crime, and his redemption. Unfortunately, as many modern critics feel, Gaskell succumbed to the preferences of her readers and publishers by recasting her plot as *Mary Barton*, the story of a virtuous working-class girl who resists the blandishments of a rich seducer and heroically saves her true love from the gallows. John Barton's melancholy story is relegated to the status of sub-plot.

One of the consequences of Mrs Gaskell's decision (which was entirely justified if she wished to reach a mass readership) was that details of John Barton's capital crime are left forever enigmatic. To summarize: driven to despair by the masters' lock-out, the trades union of which Barton is a conspiratorial ringleader resolves on an act of terror, specifically the assassination of Harry Carson, the haughty son of their chief tormentor among the employers. (Barton does not know that Harry is also the would-be seducer of his seamstress daughter, Mary.) Lots are drawn by flaring gaslight, and Barton is selected as the assassin.

The dreadful pact of the trades union occurs in chapter

16 ('Masters and Workmen'). The next chapter ('Barton's Night-Errand') is set two days later. The nature of the night-errand is ambiguous. Superficially it refers to John's being sent as a union envoy by rail to Glasgow (with a sovereign's expenses, which suggests a relatively short stay) to negotiate support from sympathetic fellow workers in the northern city. The other night-errand is, of course, the murder. In preparation John has borrowed a 'gun' from Mary's working-class suitor, Jem Wilson. A model of Smilesian self-improvement, the virtuous Wilson has nothing to do with horrible trades unions. But his deceased father and John Barton were friends, and liked to go target shooting with each other. He gives Barton the weapon (from later references to its 'stock' and the small wound it makes we apprehend that the weapon is some kind of small-bore rifle, although Mrs Gaskell is resolutely ignorant about such essentially masculine matters). Since Barton has been unemployed for some months, is starving, and addicted to opium, a sudden whim to improve his marksmanship might seem strange. But Jem is unsuspicious.

Barton, as we deduce, lays in wait for Carson in a hedge as the young man returns from his day's work in town to his father's house. One well-aimed shot to the temple does the awful deed. John Barton then disappears—possibly to Glasgow as he claimed. Although there were no witnesses, and no evident need to hurry himself, Barton leaves two clues behind at the scene of the crime. The gun is thrown down where the police can easily find it; since it has distinctive markings on it, tracing the weapon back to Wilson is the work of a few hours. John Barton also leaves behind some paper 'wadding', which he evidently removes and throws down before firing the weapon. This piece of paper has written on it the words '—ry Barton'. It was presumably placed in the barrel by Barton after cleaning the gun and hiding it in the hedgerow, prior to the murder, to protect the weapon from rusting in the dew. Although the handwriting of '—ry Barton' is distinctively Jem's, the document (a copy of Samuel Bamford's poem, 'God help the poor'; see pp. 128–9) from which the paper was torn belonged to Barton. The police, whose examination of the crime scene is perfunctory, miss this crucial piece of evidence. The paper is discovered by Mary's aunt Esther, who gives it to her niece, who destroys it. The tell-tale piece of paper persuades Mary of the dreadful fact that it must he her father who has shot Carson.

Gaskell seems to have been extraordinarily nervous of the subject of murder, and her narrative of this episode is remote and infuriatingly vague. Crucially, she does not inform us as to whether John Barton's leaving the gun and the wadding with his name on it was the consequence of panic or a deep-laid plan. Assuming the second, Barton's motives would seem to be as follows: the weapon would quickly be traced back through Jem to him, and the wadding—with the murderer's name on it—would clinch the matter. The '—ry Barton' inscription suggests a man leaving a deliberate trail for his pursuers. The chance of the premeditating, cold-blooded murderer 'accidentally' leaving a scrap of paper with his own surname by his victim exceeds even Mrs Gaskell's penchant for providential coincidence. Speculating further, we may suppose that in Glasgow—where it will cause least trauma for his daughter—Barton intends to kill himself (one of Mrs Gaskell's notes for this section mentions that Barton's 'temptation was suicide'; p. 469). Alternatively, if his nerve fails him, he may emigrate to the United States from the Scottish port under a false name, never to be seen again by his friends and family.

Barton does not know that Jem (having been informed by Esther of the other young man's evil intentions) has quarrelled publicly with Harry over Mary, and has threatened the mill-owner with dreadful consequences if he does not leave her alone. The threats are on record with the police, who—having traced the gun—arrest Jem and look no further for the assassin. On his part, Jem knows that John Barton must be the killer, since he borrowed the murder weapon a couple of days before, but out of love for Mary (whom he supposes not to know about her father's guilt) says nothing to the police or his defence lawyers, prepared as he is to go to the gallows to spare her shame. As it happens, thanks to Esther's giving her the wadding, Mary *does* know that her father is the killer. In her notes for the novel, Mrs Gaskell initially intended that Mary should visit Jem in prison; this scene was never written, presumably so that the two young people should remain at cross-purposes during the trial. Clearly, if Jem knew that Mary knew, his supreme self-sacrifice would be unnecessary. And if she knew that he knew, the sensible thing would be to instruct him to inform on John Barton rather than going, as she does, to frantic lengths to procure him an alibi.

What, meanwhile, is John Barton doing? The novel does not tell us. In a note, Gaskell wrote 'he had not heard of Jem's arrest and trial till it was over' (p. 469). This is very unconvincing. Political assassination was a sensational event (Mrs Gaskell was aware of the murder of a mill-owner; Thomas Ashton, during a strike in 1831, and admitted that it may have 'unconsciously' inspired her novel; p. ix). The *Manchester Guardian* would have been full of the matter. It is incredible that, having murdered someone, Barton would not, if he were in Glasgow, examine the papers for reports of the event. One assumes that he was not in Glasgow, nor in any city, but somewhere in the countryside trying, unsuccessfully, to screw himself up to suicide.

There is another plausible possibility, namely, that Barton intended to frame Jem. In the novel, as written, this is rendered improbable by his not knowing that Harry was the putative seducer of his daughter, and that Jem had made death-threats to his rival. In Gaskell's notes for the novel, however, Esther tells not Jem about Harry's evil plans but her brother-in-law, John. In the author's mind, then, was a pre-existing narrative in which John Barton was fully aware of how suspicious it would be if a gun belonging to Jem and a piece of paper with the words '[Ma]ry Barton' were found near the corpse. Naturally, the police would assume lover's jealousy. Suspicion would be averted from the trades union, who would have committed the perfect crime. In this version of the plot, John Barton would be many times more culpable than in the novel as written. And this second scenario would make more sense of John Barton's not hurrying back to prevent an innocent man going to the gallows in his stead (as would seem logical in the novel as written). That Jem should hang in his place was what John Barton intended from the first.

Mrs Gaskell never enlightens us. John Barton returns after the trial. Where he has been, and what he knew while he was away, we are not told. On his return we are informed that he does appear strangely crushed—'beaten down by some inward storm, he seemed to grovel along, all self-respect lost and gone.' This seems something other than assassin's remorse. It is plausible that he hates himself for having framed an innocent man; this it is that gnaws at his self-respect, and reduces him to a grovelling thing. But, as far as one can make out, Gaskell leaves the point unclear. John Barton's actions and motives are clouded in the tremendous religiosity of his final reconciliation with Carson senior and his euthanasia, forgiven by the man he has so horribly wronged.

Dives and Lazarus, man and master, are united in Christ. Nevertheless, in our speculations about the unwritten novel, 'John Barton', one is left wondering just how guilty a felon Mrs Gaskell originally had in mind.

The Oxford World's Classics *Mary Barton* is edited by Edgar Wright.

W. M. Thackeray · *Henry Esmond*

On a gross anachronism

No one much liked the ending of *Esmond* when it was published in 1852. George Eliot's comment is typical: 'the most uncomfortable book you can imagine . . . the hero is in love with the daughter all through the book, and marries the mother at the end.'[1] Eliot's discomfort has been shared by many—perhaps most—readers. It is troubling that a woman whom the orphaned hero clearly regards for much of the narrative as his mother should end up the mother of his own child. But Eliot is wrong, or impercipient, in her implication that there is something unexpected, or unprepared for, in Esmond's eventual union with Rachel. For one thing, her name, with its allusion to the long-waited-for biblical bride (Genesis 29), is a clear hint sown in the earliest pages of the narrative. At other points Thackeray is at elaborate pains to predict the final marriage, and cue the reader that Harry's ultimate happiness will lie not with the flighty Beatrix, but with her serene mother.

One such cue is found in Volume 3, Chapter 4, 'Beatrix's New Suitor' (i.e., the Duke of Hamilton), where the young heroine upbraids her discomfited 'knight of the rueful countenance' (i.e., Esmond):

> 'I intend to live to be a hundred, and to go to ten thousand routs and balls, and to play cards every night of my life till the year eighteen hundred [it is currently 1712]. And I like to be the first of my company, sir; and I like flattery and compliments, and you give me none; and I like to be made to laugh, sir, and who's to laugh at *your* dismal face, I should like to know; and I like a

coach-and-six or a coach-and-eight; and I like diamonds, and a new gown every week; and people to say—"That's the duchess—How well her grace looks—Make way for Madame l'Ambassadrice d'Angleterre—Call her excellency's people"—that's what I like. And as for you, you want a woman to bring your slippers and cap, and to sit at your feet, and cry "0 caro! 0 bravo!" whilst you read your Shakespeares, and Miltons, and stuff. Mamma would have been the wife for you, had you been a little older, though you look ten years older than she does—you do, you glum-faced, blue-bearded; little old man! You might have sat, like Darby and Joan, and flattered each other; and billed and cooed like a pair of old pigeons on a perch. I want my wings and to use them, sir.' And she spread out her beautiful arms, as if indeed she could fly off like the pretty 'Gawrie', whom the man in the story was enamoured of.

'And what will your Peter Wilkins say to your flight?' says Esmond, who never admired this fair creature more than when she rebelled and laughed at him. (pp. 363–4)

This banter ('Mamma would have been the wife for you') is a clear premonition of the final outcome, couched as it is in the elaborate games of literary allusion through which Esmond and Beatrix conduct their sexual relationship. Beatrix's mock contempt for 'your Shakespeares and Miltons' (she is, in fact, a highly civilized woman) and Harry's graceful allusion to the 'Gawries' perfectly catch the tension under its veneer of badinage. Esmond's specific reference at the end of the exchange (which may elude the modern reader, but would have been picked up by every Victorian) is to Robert Paltock's fantasia, *The Life and Adventures of Peter Wilkins*. The Gawries whom he alludes to were flying women and—significantly—the pedestrian Peter finally marries one. Henry Esmond still hopes he may one day capture the high-flying Beatrix.

It's an elegant exchange, and in keeping with the highly wrought literary quality of this section of *Esmond's* narrative. The preceding chapter, 'A Paper out of the

Spectator', is universally acclaimed as one of the most brilliant *tours de force* in Thackeray's prose, containing as it does a brilliant pastiche of Addison. Like the Peter Wilkins' allusion, Henry's mock *Spectator* essay is loaded with clever and coded messages. It follows on from his sentimental comedy *The Faithful Fool*, with its representation of the hero as Eugenio (the name is a hint to Beatrix—if she could but catch it—that Esmond is, in fact, not illegitimate but well-born, and the true Marquis; something that she will discover later to her astonishment.) The *Spectator* paper is pseudonymously addressed by 'Oedipus' to 'Jocasta', signalling that Thackeray was well aware of what knowing post-Freudian readers would find in his novel, a hundred years later. Has Esmond chosen these names unconsciously, or is he aware of certain unusual sexual desires in himself?

There is, however, a big problem with the 'Gawrie' reference in the following chapter. *The Life and Adventures of Peter Wilkins* was first published in 1751. Paltock's fable was hugely popular all through the late eighteenth and nineteenth centuries.[2] Its date (or rough period) of publication would have been familiar to any of *Esmond's* early readers capable of enjoying the *Spectator* joke. And any reader incapable of picking up that literary joke would have been beneath the narrator's notice. The 1751 date of *Peter Wilkins's* publication is, however, wholly irreconcilable with the date at which the exchange between Harry and Beatrix takes place. Thackeray is insistent about the precise year (1712) and even the month and day on which the conversation occurs. (1 April—April Fool's Day—is the date carried by the spoof *Spectator* paper). The nobleman Beatrix is about to marry (her 'new suitor', the fourth Duke of Hamilton, a historical figure) was killed in a duel in November 1712—an episode which is used climactically soon afterwards in the novel. Queen Anne enters the novel in person, further clinching the 1712–14 date of action. The third volume of *Esmond's* narrative is locked into precise, obtrusive, chronological markers which Thackeray seems intent on hammering home. In the final chapters, Esmond and Rachel emigrate to Virginia in 1718. She dies in 1736, and he is dead (presumably in the 1740s, around the period of the Scottish Rebellion) well before the publication of Paltock's work from which he is supposed to quote in 1712.

Thackeray often makes small errors of chronology But the 'Gawrie' lapse is gross—suspiciously so. It is equivalent to Hardy's Jude quoting at length from *The Waste Land*, or Emma Woodhouse comparing Harriet Smith's plight to that of Jane Eyre. It jars. And it occurs at a section in the novel where Thackeray is so exact and virtuosic in his play with eighteenth-century comic literature (a subject on which he was currently lecturing to British and American audiences, and on which he was the leading authority)[3] as to render it wholly incredible that he did not register the error. And—if by some blindspot he did not register it—why did he not remove the reference in the revised edition of *Esmond* (as he dropped other embarrassing material)?

Readers certainly registered the lapse. Thackeray's 'Peter Wilkins' solecism was gleefully seized on by early contributors to *Notes and Queries*, and every edited text of the novel ruefully notes it as a damaging anachronism. Rather than join the querists and annotators in gloating over or lamenting what looks like a monumental blooper, it makes better sense to assume that Thackeray put this anachronism into his text, and kept it there in revised editions, for an artistic purpose. That purpose can be deduced from the intricate narrative framework of the novel. The story is 'edited' by the daughter of Rachel and Esmond, Rachel Esmond Warrington ('REW'), who has had the written 'Memoirs' (i.e. the text of the novel) from Esmond's own pen. Her garrulous and frequently vulgar preface is dated 3 November 1778.

'REW' is an unequivocally ridiculous figure as she appears on the edges of the story: vain and jealous of her mother's claim on her father's affections, vengeful against 'Mrs Tusher' (i.e. Beatrix—who has to descend a long way from her 'Duchess' aspirations), headstrong, and rather stupid. And in the last volume REW loses editorial control over her material, inserting a series of increasingly fatuous footnotes, culminating in a bizarre footnoted dissertation in Volume 3, Chapter 10 on her father's Christian gentleness of manners to all and sundry, not excluding 'the humblest negresses on his estate' (p. 431).

What we are led to assume is that REW's spoiling pen has also intervened in the text of the 'Memoirs', as well as at its prefatory and marginal edges. The passages which speckle the third volume, predicting the eventual 'Darby and Joan' happiness of Esmond and Rachel have been inserted by REW we may plausibly deduce, at some point after 1751 and her parents' death. Like other pious biographers, she wants to project a whitewashed image of her parents to the world. Thackeray himself was morbidly aware of the dangers of such loyal filial impulses, and the kind of pious absurdities they might lead to. He frequently instructed his daughters over the last years of his life, 'No biography!'—an instruction which his daughter Anne religiously observed (there was no family-authorized biography of Thackeray until the 1950s, when Gordon Ray was finally given permission).[4] REW is not restrained as strictly as Thackeray's daughters on the question of paternal biography—indeed, Esmond seems to have consciously charged his daughter with the care of his 'Memoirs' and a *nihil obstat* on their eventual publication. REW as we can see, habitually exceeds her strict editorial remit, adding her own 'improvements' which go well beyond the signed footnotes into the actual text of Esmond's memoirs. But, impulsive and unscholarly as she is, REW betrays her interference by a gross solecism—a solecism so gross, indeed, that no one who knows the author(s) can picture Thackeray or Esmond making it. It is only too easy, however, to imagine Rachel Esmond Warrington perpetrating the blunder.

The Oxford World's Classics *The History of Henry Esmond* is edited by Donald Hawes.

Charles Dickens · *Bleak House*

What is Jo Sweeping?

In Philadelphia, in the winter of 1994–5, the pedestrian waiting at any of the cross-streets at Center City (as the heart of the metropolis is called) would find a helpful person with a broom sweeping an already spotless and carefully built-up gutter. Since the winter was the warmest since the thermometer was invented and the crossings no dirtier than the pavements, many tourists must have wondered why the governors of a city in financial crisis should have supplied this service. The reason for the army of crossing sweepers was the series of devastating ice-storms which had raged through the city in 1993–4, and the crop of lawsuits brought against Philadelphia by injured pedestrians.

Why, one may go on to wonder, did the Victorians need crossing-sweepers of the kind who furnish main characters in at least three novels of the period? (Thackeray's *The Yellowplush Papers*, 1841; Bulwer-Lytton's *Lucretia*, 1846; and—most famously—*Bleak House*, 1853.) If Dickens's Jo is a sweeper, what precisely does he sweep? An answer is given in the first paragraph of the novel:

> London. Michaelmas Term lately over, and the Lord Chancellor sitting in Lincoln's Inn Hall. Implacable November weather. As much mud in the streets, as if the waters had but newly retired from the face of the earth, and it would not be wonderful to meet a Megalosaurus, forty feet long or so, waddling like an elephantine lizard up Holborn Hill.

'Mud', it would seem, is what the mid-Victorian sweepers swept.[1] Here Dickens pictures it as the primal soup from which the post-diluvian, fallen world, has evolved. But, as he must have pronounced it, the word 'mud' had overtones of another of Dickens's favourite words, *merde*—shit. *Merde* was memorably compounded elsewhere into two of the author's most biting names, Merdle and Murdstone. The anglicized form of the word is picked up by T. S. Eliot, in *Gerontion*, in one of his more Dickensian visions of urban life:

> The goat coughs at night in the field overhead;
> Rocks, moss, stonecrop, iron, merds.

As Fred Schwarzbach points out, in his book *Dickens and the City*, mud and *merde* had much in common in 1853:

> The mud of mid-century London was, after all, quite different from the harmless if messy stuff children today make into pies. It was compounded of loose soil to be sure, but also of a great deal more, including soot and ashes and street litter, and the fecal matter of the legion horses on whom all transport in London depended. In addition, many sewers (such as they were) were completely open, and in rainy weather would simply overflow into the streets. Dogs, cattle in transit either to Smithfield or through the town (many dairies were still inside the city), and many people as well used the public streets as a privy, but then even most privies were simply holes in the ground with drainage into ditches or another part of the street. (London was still a good fifteen years away from having an effective drainage system.) The mud must at times have been nothing less than liquid ordure.[2]

Norman Gash, in *Robert Surtees and Early Victorian Society* is more analytic, if no less disgusted:

> Everyone agreed that London was the dirtiest of English cities. The central streets were covered with a thick layer of filth, the principal ingredients of which were horse manure, stonedust from the constant grinding of iron tyres on the paving stones, and soot descending from the forest of London chimneys. The resultant compound resembled a black paste, which clung

glutinously to everything it touched and emitted a characteristic odour reminiscent of a cattle-market. Though contractors were employed to clean the streets, the work was hard and ill-paid; it could only be done at night; and the filth carried away had little commercial value. London produced in fact such a vast amount of excrement, both animal and human, that the surrounding agricultural districts were unable to absorb it. As a result the streets seemed to remain permanently foul. To get across busy streets, especially in wet weather, was a hazardous business except at the regular paths—'isthmuses of comparative dry land', as one observer called them—kept clean by the unremitting work of crossing-sweepers.[3]

Henry Mayhew devotes a fairly long section of *London Labour and the London Poor* (1851) to crossing-sweepers—'a large class of the Metropolitian poor', as he terms them. Mayhew noted: 'We can scarcely walk along a street of any extent, or pass through a square of the least pretensions to "gentility", without meeting one or more of these private scavengers.'[4] As Mayhew observed, for most of these broom-wielders sweeping was an excuse for begging. Using this cover was attractive because: (1) it required very little capital to commence the business; (2) the pseudo-occupation enabled the sweeper to solicit gratuities without being considered in the light of a street beggar; (3) being seen in the same place constantly excited the sympathy of the neighbouring householders, encouraging the donations of small weekly allowances or 'pensions'. The Victorian crossing-sweeper was exactly analogous to the ubiquitous windscreen cleaner to be found importuning motorists at London and New York traffic-lights in the 1980s and 1990s. As is often observed, one of the finest achievements of Reagonomics and Thatcherism was to reinvent the able-bodied pauper and the Victorian street beggar. Dickens would have been amazed.

As is his wont, Mayhew anatomizes the different classes

of crossing-sweeper in great factual detail. The average income is, he reckons, a shilling a-day. The sweepers, he judges, constitute 'the most honest of the London poor' (p. 466). They vary from the well-known London landmarks, who have occupied the same pitch for years, to the 'irregular' bands of young girl and boy crossing-sweepers. Mayhew makes the chilling point that the children sweepers can only survive for any length of time in the streets by forming themselves into organized gangs, living and working together co-operatively. As he notes:

> The irregular sweepers mostly consist of boys and girls who have formed themselves into a kind of company, and come to an agreement to work together on the same crossings. The principal resort of these is about Trafalgar Square, where they have seized upon some three or four crossings, which they visit from time to time in the course of the day. (ii. 466)

Jo, one recalls from Chapter 11, has his pitch by St Martin-in-the-Fields Church, in Trafalgar Square. And the reason that he is 'moved on', so calamitously, is because he is alone—he has not, that is, acquired sufficient street smarts to realize that he can only survive if he throws in his lot with others who hunt in packs.

One of the modern reader's principal problems in visualizing the world of *Bleak House* is not being sure of how far one should let one's imagination rip. Should we accept Fred Schwarzbach's Dantean vision of a London swimming in 'liquid ordure'? Or were the urchins around Trafalgar Square simply going through the motions (like windscreen cleaners at modern crossroads), the more conveniently to beg a penny or two from passers-by? And even if there were a lot of droppings in the street, would they necessarily be so obnoxious? George Orwell (in his essay on Swift) makes the point that horse dung—given the beast's graminivorous diet and fast metabolism—is not, as dung goes, physically nauseating (compared to, say, the tons of dog-mess dropped daily on London parks today). The streets of the West End in 1852 might not have been more unpleasant than, say, the stretch of Oxford Street outside McDonalds with its detritus of French-fry packets, milk-shake cartons, and hamburger grease. A modern-day Jo would be rather welcome sweeping the crossroads at Tottenham Court Road and Oxford Street.

This question relates to another quarrel in recent Dickens studies—namely, what constitutes the 'dust' in the dust heaps in *Our Mutual Friend* (1865). Humphry House, in his influential study *The Dickens World* (1941), announced to a less squeamish age than the Victorians that 'dust' was a euphemism for 'human excrement'. The revelation gave a new *frisson* to such descriptions as that of Silas Wegg, stumping triumphantly over the mounds, his wooden leg puncturing the surface crust, leaving a trail of mephitic vapours behind him.

It is true that the English habitually use the word 'dust' euphemistically—as in 'dustman' and 'dustbin'. But in his 1971 Penguin Classics edition of *Our Mutual Friend*, Stephen Gill disputes House's garish vision of a mountain of decaying human turd lowering over the love-making of Bella and Harmon. Drawing on Mayhew, Gill argues that human excrement was too liquid to form into heaps and too valuable to dispose of as garbage. The dust heaps, as Mayhew pointed out, were composed of seven elements, all of them valuable as salvage, but none so valuable as human excrement (which was used as manure). They were: (1) fine dust, sold to brickmakers; (2) cinders, or brieze, sold to brickmakers for breeze-blocks; (3) rags, bones, and old metal; (4) old tin and iron; (5) old bricks and oyster shells, sold to builders for footings; (6) old boots and shoes; (7) money and jewellery. It has been pointed out elsewhere, however, that Mayhew does indicate that another main constituent of the heaps was manure—

which would suggest that animal (as opposed to human) excrement formed a large part of the substance.[5] The question is still somewhat open.

Decoding Dickens on the mud/*merde*, dust/shit question is difficult, because in general the Victorian novelists (excluding the pornographers) are extraordinarily fastidious about mentioning human or animal waste—excretion, like sex, is a rigorously tabooed subject. In the whole of Surtees, for instance, one detects few clear references to horse dung, although a moment's reflection reveals that the Jorrocks narratives must be ankle-deep in the odoriferous substance (but then, how many horse droppings does one see in John Wayne cowboy movies?). Of course, one can often supply what the novelist carefully omits to name directly. In the following passage by Mrs Gaskell from *Mary Barton* describing the utter squalor of a Manchester slum dwelling it is not hard to fill in the lexical blank of 'slops of *every* description':

> [The street] was unpaved; and down the middle a gutter forced its way, every now and then forming pools in the holes with which the street abounded. Never was the old Edinburgh cry of 'Gardez l'eau!' more necessary than in this street. As they passed, women from their doors tossed household slops of *every* description into the gutter; they ran into the next pool, which overflowed and stagnated. Heaps of ashes were the stepping-stones, on which the passer-by, who cared in the least for cleanliness, took care not to put his foot. (p. 66)

As the World's Classics editor, Edgar Wright, points out, the Edinburgh reference is to the emptying of chamber-pots, and 'ashes' is to be read 'as a euphemism for excrement'. The experienced reader of Victorian fiction becomes adept in filling in such blanks and picking up such hints. It is, for instance, a commonplace that characters in Victorian fiction do not go to the lavatory. Yet, if we are moderately sensitive, we apprehend that when Casaubon asks Dorothea and Celia twice during their first visit to his house at Lowick if they are 'tired' he is politely inquiring if they wish his maid-servant to conduct them to the ladies' room.[6]

It is also the case that a writer like Dickens did not have to specify what London street mud was like for the reader of his time, any more than he was obliged to record that the same streets were cobbled or that the carriages were pulled by horses. Everybody already knew. On his part Mayhew, despite the most pedantic anatomization of the condition and variety of street-sweepers, rarely feels obliged to allude to what is actually being swept. He did not have to. The reader would only have to look from the page to the sole of his shoe. In a hundred years' time students may wonder exactly what kind of slime, grime, grit, or filth accumulated on the windscreens of late-twentieth-century cars that needed to be cleaned off so frequently. No one, writing today in newspapers, would feel obliged to describe the exact composition of the grime thrown up on our roads (sprayed mud, squashed insects, a patina of oil and rubber). It may well require a footnote a century hence. And casual readers, a century hence, might well surmise that city streets in the 1990s were exceptionally dirty—otherwise why was there any need for that army of windscreen cleaners?

With Dickens, however, there is a peculiarly Dickensian problem. His disinclination to specify mud/*merde*, dust/shit, makes it difficult for the historically alienated modern reader to know whether or not the writer is indulging in his characteristic hyperbole, and to what degree. Clearly the megalosaurus in the opening passage of *Bleak House* is a flight of hyperbolic fancy (inspired, I would guess, by the papier-mâché dinosaurs constructed for the Crystal Palace Exhibition, a couple of years earlier). But, if we follow Schwarzbach, Dickens's description of the street mire in Holborn is, if anything, understated—'mud' is not hyperbole, but litotes.

The point is clearer if one goes on to the parallel scene in Chesney Wold, in the opening of the second chapter. Here the elemental excess is not earth/mud, but water/rain:

> The waters are out in Lincolnshire. An arch of the bridge in the park has been sapped and sopped away. The adjacent low-lying ground, for half a mile in breadth, is a stagnant river, with melancholy trees for islands in it, and a surface punctured all over, all day long, with falling rain. My Lady Dedlock's 'place' has been extremely dreary. The weather, for many a day and night, has been so wet that the trees seem wet through, and the soft loppings and prunings of the woodman's axe can make no crash or crackle as they fall. The deer, looking soaked, leave quagmires, where they pass. The shot of a rifle loses its sharpness in the moist air, and its smoke moves in a tardy little cloud towards the green rise, coppice-topped, that makes a background for the falling rain.

Unlike London mud, English rain has not altered in composition over the last hundred years, and one can precisely gauge how far Dickens is over-writing here, in his description of the unnaturally silent landscape. But is the description of the excess of mud in Holborn hyperbolic? And for mud, should we read *merde*? One is driven to surmise. My guess is that the crossing Jo superintends at Trafalgar Square is not all that bestrewn with muck, and although by no means wholesome is no more filthy than the same streets today at the end of a busy week-day. What, one may ask oneself could an undersized boy, with a worn-out broom (such as is described by Dickens and pictured below by Frank Beard) be expected to clean? Not much. Is the gentleman depicted complacently handing over his coin really worried about getting his shoes and understrapped trousers dirty? Or is discreet charity his main motive? Certainly the artist is careful to sketch in two carriages in the background to the picture, and there will be droppings. But not, one apprehends, the ankle-deep tide of filth which recent commentary has pictured swirling around the public places of Victorian London.

The Oxford World's Classics *Bleak House* is edited by Stephen Gill. The Oxford World's Classics *Mary Barton* is edited by Edgar Wright.

The Boy Crossing-Sweepers

Charlotte Brontë · *Villette*

Villette's double ending

Critics have traditionally been fascinated by the enigmatic ending of *Villette*—particularly hyper-modern critics who see in the novel an anticipation of the 'problematized text', so beloved of deconstructionists and of theorists generally.[1] To summarize: at the end of the narrative, Lucy Snowe has her virtuous pluck rewarded by the declared love and marriage proposals of her stern 'professor', Paul Emanuel. But before he can make Lucy his wife, Paul must spend three years working in the French protectorate Guadaloupe. The reasons for his exile to this far-off place are vaguely communicated to the reader by Lucy, in Chapter 39: 'its alpha is Mammon, and its omega Interest' (p. 461), she declares. Madame Walravens, we are informed, has earlier inherited by marriage a large estate at Basseterre, on the West Indian island: 'if duly looked after by a competent agent of integrity' for 'a few years', the estate will be 'largely productive'. Madame Walravens asks Paul to be her 'competent agent'. As Lucy observes, such a wish is a command: 'No living being ever humbly laid his advantage at M. Emanuel's feet, or confidingly put it into his hands, that he spurned the trust, or repulsed the repository' (p. 462). Whatever might be Paul's 'private pain or inward reluctance to leave Europe', he accedes.

It is, on the face of it, strange that Paul should accede. The claims of Lucy and his own happiness would seem to be stronger than the financial convenience of Madame Walravens. He is not, in any case, a businessman, but a schoolteacher and a very good one—if a little too fond of 'discipline'. But it is not hard to deduce what duties Paul Emanuel is required for. The date of *Villette's* action is the early 1840s (the period of Charlotte Brontë's own residence in Brussels, 1842–4). Slavery had finally been abolished in the British West Indies in 1833, precipitating a disastrous collapse in the sugar industry, with the widespread defection of pressed labour. In neighbouring Guadaloupe, under the unenlightened French imperial regime, the institution of slavery (and the profitability of the sugar plantations) was to limp along until its eventual, long-overdue abolition in 1848. The stern and dictatorial Professor Emanuel—the bully of Madame Beck's classroom—has been recruited to rally the increasingly dissident slave labourers of Madame Walravens's estate, with whips and scorpions, if necessary. For 'competent', read 'brutal'. There is another putative factor in the virtuous lady's choosing Monsieur Emanuel as her overseer. He has shown himself, in his attendance at Madame Beck's establishment, remarkably capable of restraining himself sexually in the presence of nubile young women. All nineteenth-century accounts of Guadaloupe stress that it is a place of almost irresistible temptation for European males. As the *Encyclopaedia Britannica* (14th edition, 1929) records:

> Guadaloupe has a few white officials and planters, a few East Indian immigrants from the French possessions in India, and the rest negroes and mulattoes. These mulattoes are famous for their grace and beauty of both form and feature. Women greatly outnumber men, and illegitimate births are very numerous. (x. 927)

Clearly only a man of iron self-discipline can be trusted in such a Sodom.

One assumes that Paul's motives for exiling himself from Lucy are at least partly to test his bride-to-be, to try her ability to survive without him. Is she worthy to be

Madame Emanuel? He is fond of imposing such ordeals. In the three years of his absence Lucy must prove her worth by setting up a school. She succeeds magnificently, inspired by the lessons in discipline and self-discipline that she has learned from her professor.

The relationship between the lovers during Paul's absence is sustained by passionate letters. The novel concludes with a coda which switches dramatically from the past to the present tense: 'And now the three years are past. M. Emanuel's return is fixed.' It was early summer when he left (the roses were in bloom). Now, on the eve of his return, it is autumn and the season of equinoctial storms. Lucy apostrophizes the elements as her demonic foe:

> The wind shifts to the west. Peace, peace, Banshee—'keening' at every window! It will rise—it will swell—it shrieks out long: wander as I may through the house this night, I cannot lull the blast. The advancing hours make it strong: by midnight, all sleepless watchers hear and fear a wild south-west storm. (p. 495)

A cataclysmic storm duly rages for seven days, an appropriately de-creating span of time which will, we fear, return Lucy's universe to chaos. The novel concludes with an enigmatic and emotionally exhausted last two paragraphs:

> Here pause: pause at once. There is enough said. Trouble no quiet, kind heart; leave sunny imaginations hope. Let it be theirs to conceive the delight of joy born again fresh out of great terror, the rapture of rescue from peril, the wondrous reprieve from dread, the fruition of return. Let them picture union and a happy succeeding life.
>
> Madame Beck prospered all the days of her life; so did Père Silas; Madame Walravens fulfilled her ninetieth year before she died. Farewell. (p. 496)

Does Emanuel drown or does he survive drowning? The reference to Madame Walravens indicates that Lucy is writing many years after the event, so the outcome must be known—despite the present tense used to evoke the storm.

The novel's internal structure of allusion is enigmatic on the question. The 'banshee' reference looks all the way back to Chapter 4, which features a terrible storm while Lucy is in England, in the service of the invalid Miss Marchmont. That storm is described as being accompanied by a 'subtle screeching cry'. Looking forward, beyond the events of Villette's narrative, Lucy records that:

> Three times in the course of my life, events had taught me that these strange accents in the storm—this restless, hopeless cry—denote a coming state of the atmosphere unpropitious to life. Epidemic diseases, I believed, were often heralded by a gasping, sobbing, tormented long-lamenting east wind. Hence, I inferred, arose the legend of the Banshee. (p. 38)

Miss Marchmont dies that night.

The second of the three occasions alluded to by Lucy in the above passage is recorded in Chapter 15. The heroine is now in the less congenial service of Madame Beck. She has been subject to 'peculiarly agonizing depression' (p. 159) and feverish delirium. Recovered from her nightmares, but still weak, she arises from her bed, unable to bear the 'solitude and the stillness' of the dormitory any longer. It is evening, and the darkening sky is terrible with the threat of storm:

> from the lattice I saw coming night-clouds trailing low like banners drooping. It seemed to me that at this hour there was affection and sorrow in Heaven above for all pain suffered on earth beneath; the weight of my dreadful dream became alleviated. (p. 160)

Lucy unwisely ventures out into the storm in her weakened, semi-invalid state. The end of this chapter marks the gap between the first and second volumes of the original three-volume edition brought out by Smith, Elder & Co in January 1853. For the mass of circulating-library readers, rationed by their one-guinea subscriptions to one volume at a time, this gap would entail more than simply reaching for the next volume. There would, probably, be a longish interval during which volume 1 was returned (possibly after a delay while some other member of the family read it) and the second volume borrowed from the library (possibly after yet another delay if one had to visit Mudie's main establishment in Bloomsbury, or if all the second volumes were 'out'). Readers, thus kept in suspense, might reasonably expect that Lucy was about to die. Chapter 15 and volume 1 end with the dramatic statement: 'I seemed to pitch headlong down an abyss. I remember no more.'

Is Lucy dead? Almost dead, it transpires. The opening of the second volume (Chapter 16) picks up on the 'abyss' reference: 'Where my soul went during that swoon I cannot tell.' Lucy has experienced, it seems, an after-death experience:

> Whatever [my soul] saw, or wherever she travelled in her trance on that strange night, she kept her own secret; never whispering a word to Memory, and baffling Imagination by an indissoluble silence. She may have gone upward, and come in sight of her eternal home, hoping for leave to rest now, and deeming that her painful union with matter was at last dissolved. While she so deemed, an angel may have warned her away from heaven's threshold, and, guiding her weeping down, have bound her, once more, all shuddering and unwilling, to that poor frame, cold and wasted, of whose companionship she was grown more than weary. (p. 165)

Miss Soul's return (that is to say, Lucy's revival) is further metaphorized in terms of a painful rescue from drowning. This, one assumes, is the second 'banshee' experience in which the victim is pulled back from the very jaws of death. The third banshee-storm-death episode is that connected with Paul Emanuel's return voyage from Guadaloupe and his shipwreck at sea. Does it betoken death (as with Miss Marchmont) or a terrifying brush with death (as with Lucy)?

Charlotte Brontë evidently received some querulous correspondence on the subject of her indeterminate ending. As Mrs Gaskell records, two of the author's female contemporaries wrote demanding 'exact and authentic information respecting the fate of M. Paul Emanuel'. Brontë wrote to her editor at Smith, Elder & Co. saying she had dispatched an answer, 'so worded as to leave the matter pretty much where it was. Since the little puzzle amuses the ladies, it would be a pity to spoil their sport by giving them the key.'

As Margaret Smith and Herbert Rosengarten suggest, in their World's Classics edition of *Villette*, we should also consider in this context Brontë's sharper letter to her publisher, George Smith, on 26 March 1853:

> With regard to that momentous point—M. Paul's fate—in case anyone in future should request to be enlightened thereon—they may be told that it was designed that every reader should settle the catastrophe for himself, according to the quality of his disposition, the tender or remorseful impulse of his nature. Drowning and Matrimony are the fearful alternatives. The Merciful . . . will of course choose the former and milder doom—drown him to put him out of pain. The cruel-hearted will on the contrary pitilessly impale him on the second horn of the dilemma—marrying him without ruth or compunction to that—person—that—that—individual—'Lucy Snowe'.

Charlotte Brontë supplies us in this sarcastic letter with the 'key to the puzzle'. That is to say, one could only sustain the 'sunny' reading of the novel's ending if one equated the disaster of Paul's drowning with the 'disaster' of his marrying the woman he loves and who loves him. It is 'moral' for Lucy to prevaricate if Paul Emanuel has indeed died at sea, so as not to discomfit her less emotionally sturdy readers. Hiding her misery is a brave and admirable thing to do. Had Paul Emanuel returned, it would have been reprehensible to have disguised or withheld the fact. Such behaviour could only be construed as a claim for undeserved pity and sympathy.

There is, in short, no problem with the conclusion of *Villette*, if one gives it a moment's thought. Paul Emanuel drowns—end of story (literally). Mrs Gaskell confirms the point by reference to privileged family testimony:

> Mr Brontë was anxious that her new tale should end well, as he disliked novels which left a melancholy impression upon the mind; and he requested her to make her hero and heroine (like the heroes and heroines in fairy-tales) 'marry, and live happily ever after'. But the idea of M. Paul Emanuel's death at sea was stamped on her imagination, till it assumed the distinct force of reality; and she could no more alter her fictitious ending than if they had been facts which she was relating. All she could do in compliance with her father's wish was so to veil the fate in oracular words, as to leave it to the character and discernment of her readers to interpret her meaning. (p. 538)

The interpretation is simple enough, and few sensible readers of the novel can have given credence to the spurious 'sunny' ending.[2] More interesting—particularly in the context of 1853—is the device of the double ending, the reader being left free to choose between a 'real' or a 'fairy-tale' version. *Villette* came out in January 1853. In October of the same year Thackeray began to serialize *The Newcomes.* This massive saga-novel was to continue as a monthly serial for the following twenty-three months, until August 1855. The main strand of Thackeray's narrative deals with the careers of the cousins Ethel and Clive Newcome, their true love for each other, and their respective unhappy marriages to less-congenial partners.

The Newcomes ends with one of the greatest effusions of

pathos in Victorian fiction, the death of Colonel Newcome, and Clive and Ethel aching for each other but forever separated. There follows a coda in which Thackeray, apparently *in propria persona*, declares that: 'Two years ago, walking with my children in some pleasant fields, near to Berne, in Switzerland, I strayed from them into a little wood: and, coming out of it presently, told them how the story had been revealed to me somehow, which for three-and-twenty months the reader has been pleased to follow' (p. 1007).

Thackeray then embarks on an extended fantasia about what his characters are doing in 'Fable-land'. His belief is, he says, 'that in Fable-land somewhere Ethel and Clive are living most comfortably together'. That is, as man and wife. 'You', Thackeray tells his reader:

> may settle your fable-land in your own fashion. Anything you like happens in fable-land. Wicked folks die à propos . . . annoying folks are got out of the way; the poor are rewarded—the upstarts are set down in fable-land . . . the poet of fable-land . . . makes the hero and heroine happy at last, and happy ever after. Ah, happy, harmless fable-land, where these things are! Friendly reader! may you and the author meet there on some future day! He hopes so; as he yet keeps a lingering hold of your hand, and bids you farewell with a kind heart. (p. 1009)

It seems an uncannily close echo of the Revd Brontë's demand that his daughter's novel should deny its own character, and conclude with a 'fairy-tale' ending. And there is, as it happens, a graphic record of just such pressure being brought to bear on Thackeray. While lecturing in Coventry during the course of the serial run of *The Newcomes*, Thackeray was entertained by the Brays and Hennells (George Eliot's friends). Miss Hennell, as the ladies' spokeswoman, said:

> 'Mr Thackeray, we want you to let Clive marry Ethel. Do let them be happy.' He was surprised at their interest in his characters

[and replied] 'The characters once created *lead me*, and I follow where they direct.'[3]

None the less, in his fable-land coda, as had Charlotte Brontë a couple of years earlier, Thackeray threw a sop to his soft-hearted readers and Coventry's imperious need for 'sunny' endings.

In April 1854, just over a year after *Villette's* publication and a good year before the end of *The Newcomes'* run, Dickens began to serialize his new novel *Hard Times* in *Household Words*, where it ran weekly until August. *Hard Times* finishes with a visionary coda in which the narrator pictures a series of happy endings, including that of Louisa, released from Bounderby and

> again a wife—a mother—lovingly watchful of her children, ever careful that they should have a childhood of the mind no less than a childhood of the body, as knowing it to be even a more beautiful thing, and a possession, any hoarded scrap of which, is a blessing and happiness to the wisest. (p. 397)

'Did Louisa see this?' the narrator asks: 'Such a thing was never to be.'

These three novels, by the three leading novelists of 1853–5, all employ the same striking terminal device of the double ending. One of those double endings is harshly 'realistic', and aimed at tough readers (with whom the novelist is clearly in closer sympathy). For softer-minded readers, of 'sunny' disposition, an alternative 'fairy-tale' ending is supplied (or in Dickens's case, hinted) in which Paul Emanuel returns to the embraces of his little Protestant Lucy, Clive and Ethel are united to the clashing peals of wedding bells, and Louisa recovers from her near seduction by Harthouse to become a respectably fulfilled materfamilias. 'You pays your money and you takes your pick', the novels seem to say.

One can suggest a reason for this epidemic of double

endings in 1853–5. Clearly Charlotte Brontë, Charles Dickens and Thackeray were responding to the pressure of a new reading public, one that wanted happy endings and could make its wants felt. That reading public (with its representative spokespersons like the Revd Brontë and Miss Hennell) had been massively organized and empowered by the new phenomenon of the circulating library—notably Mudie's. Charles Mudie's 'Leviathan' had started modestly enough in Southampton Row in the 1840s. In 1852, however, the firm moved into massive new premises at the corner of New Oxford Street and Museum Street. As Guinevere Griest records: 'During the ten years between 1853 and 1862, Mudie added almost 960,000 volumes to his library, nearly half of which were fiction'.[4] Mudie was suddenly the biggest bulk purchaser of new novels in the kingdom. And the proprietor of the Leviathan demanded happy endings, on behalf of his customers. As Griest again records, 'over and over again works were censured because they were "disagreeable" or "unpleasant", qualities which Mr Mudie's readers did not care to find in their novels' (p. 136).

The peremptory demands of the library reader for sunshine and their resentment of anything 'disagreeable' were beginning to be focused on the novelist in 1853–5, by direct pressure from the publisher, himself under direct pressure from the bulk-buying libraries. What the endings of these three high-profile novels of 1853–5 indicate is that on their part the novelists had registered the demands of this newly mobilized force of library readers, and were devising subtle strategies of resistance. This complex tug-of-war between novelist and the tyrannic circulating library was to continue until 1894 and the collapse of the three-decker novel under the assault of novelists like George Moore and Thomas Hardy, enraged by the constraints that Mrs Grundy (alias Mr Mudie, the 'nursemaid' of literature) were imposing on their art and their claims to the privileges of realism.

The Oxford World's Classics *Villette* is edited by Margaret Smith and Herbert Rosengarten with an introduction by Tim Dolin. The Oxford World's Classics *The Newcomes* is edited by Andrew Sanders. The Oxford World's Classics *Hard Times* is edited by Paul Schlicke.

George Eliot · *Adam Bede*

What is Hetty waiting for?

In their forthrightly entitled article, 'Victorian Women and Menstruation', Elaine and English Showalter note that:

> Few taboos evoke as forceful and as universal a response as that surrounding menstruation. Even the redoubtable Marquis de Sade, who took a prurient delight in moldy feces and decapitated dogs, appears to have regarded menstruation with faint distaste . . . Small wonder that even Victorians as open-minded as Florence Nightingale and John Stuart Mill maintain an almost complete silence on the subject. (p. 83)[1]

'Unlike sexual activities', the authors continue, 'menstruation has no literary reflection, true or false.' In the quarter of a century since the Showalters wrote their article critics have become very ingenious at uncovering the repressed consciousness of nineteenth-century fiction—most aggressively in Eve Kosofsky Sedgwick's polemic, 'Jane Austen and the Masturbating Girl'.[2] None the less, the menstruating woman has not been located and the Showalters' claim about 'no literary reflection' would seem to hold up. There is, however, one exception. Not surprisingly, it is to be found in the toughest-minded of the mid-Victorian novelists, George Eliot.

In what follows, I shall attempt to read the critical middle chapters of *Adam Bede* as a contemporary Victorian woman of the world might have read them—looking shrewdly between the lines for hints, clues, and coded references. Chapter 27, 'A Crisis', begins with specific time-markers (George Eliot is unusually precise about days, weeks, and months in this section of the narrative, a feature which might well alert our notional woman reader).[3] It is, as the crisis looms, 'beyond the middle of August'—the 18th, as we later learn. It is coming up to harvest time in Loamshire, and the local farmers are anxious that the high August winds will damage the crop of wheat, which is ripe in the ear but not as yet gathered in: 'If only the corn were not ripe enough to be blown out of the husk and scattered as untimely seed'.

Untimely seed is being spilled elsewhere. As we apprehend from hints earlier, Arthur and Hetty are meeting in the Chase for clandestine sexual assignations. Walking through these woods at sunset on 18 August, Adam comes across Arthur and Hetty kissing. In his innocence, the young carpenter supposes that the intimacy which he has accidentally witnessed has gone no further than kissing and cuddling. Adam confronts Arthur who—with secret relief—quickly discerns that Adam has not realized how far his misconduct with Hetty has gone. Furtively, the squire notes that 'Adam could still be deceived'. Deceived, that is, not as to the fact that he has stolen Hetty's heart, but that he has stolen her virginity as well. The two young men come to blows, fighting 'with the instinctive fierceness of panthers in the deepening twilight'. Adam, the stronger man, eventually knocks Arthur out.

Chapter 28 of *Adam Bede* opens with a chastened Donnithorne informing his victorious opponent that 'I'm going away on Saturday, and there will be an end of it'. Adam—who is not entirely a fool—inquires as to whether the affair is only a matter of 'trifling and flirting, as you call it' and, on being dishonestly reassured on the matter, he demands that a letter be written, disabusing Hetty of any expectation that Arthur can ever marry her. In Chapter 29 Arthur concludes that Adam '*must* be satisfied, for more reasons than one'. Arthur dimly sees as his salvation

Adam going on to marry Hetty, and never—till his dying day—finding that she has been seduced.

As George Eliot insinuates, Arthur also sees the awful prospect (among other awful prospects) that Hetty may be pregnant:

> A sudden dread here fell like a shadow across his imagination—the dread lest she should do something violent in her grief; and close upon that dread came another, which deepened the shadow. But he shook them off with the force of youth and hope. What was the ground for painting the future in that dark way? It was just as likely to be the reverse ... There was a sort of implicit confidence in him that he was really such a good fellow at bottom, Providence would not treat him harshly.

The first dread is that Hetty may kill herself. The second dread, as any wide-awake Victorian adult reader would apprehend, is that she may be with child. Trusting to his luck, as have multitudes of reckless young men before him, Arthur resolves to keep Adam in ignorance. Somewhere too, at the very back of his mind, will be the unformed idea that surely—if she does find herself in trouble—Hetty will know some old woman somewhere who will help her get rid of the unwanted thing.

Chapter 30 describes the delivery of Arthur's letter via Adam. Hetty, before reading it, is also quick to perceive that Adam does not realize how far things have gone between her and Arthur. Chapter 31, 'In Hetty's Bed Chamber', shows Hetty reading the letter that will ruin her life. It contains the awful sentence:

> I know you can never be happy except by marrying a man in your own station; and if I were to marry you now, I should only be adding to any wrong I have done, besides offending against my duty in other relations of life.

Clearly enough, Arthur is advising Hetty to marry Adam, and better do it quickly to be on the safe side. The letter finishes with the instruction, 'Do not write unless there is something I can really do for you'. He leaves an address—we infer he is still worried about the possibility of pregnancy.

As the narrator puts it, Hetty's 'short poisonous delights' (by which we may assume that the meetings in the Chase were fumbled and dubiously joyful) have 'spoiled her life forever'. Chapter 33 opens with the harvesting of the barley and the gathering of the season's apples and nuts. It is the annual harvest festival celebrated by country folk on 29 September every year: 'Michaelmas was come, with its fragrant basketfuls of purple damsons, and its paler purple daisies'. Assuming a last sexual encounter on 18 August, Hetty will still be unsure whether her own fruitfulness is to be added to that of the season. If her period is a week or two late, that could well be the effect of the shock of Arthur's letter.

Mrs Poyser notes a 'surprising improvement in Hetty'. The girl is, for a change, quiet and submissive. She is also setting her cap at Adam, which is slightly odd. Hetty is described smiling at her young suitor, 'but there was something different in her eyes, in the expression of her face, in all her movements Adam thought—something harder, older, less child-like'. In Chapter 34 we are told it is now 2 November: and on this Sunday, on the way back from church, Adam proposes to Hetty. He notices as he does so that she exhibits of late 'a more luxuriant womanliness'. This we may gloss as a fuller physical shape—particularly around the bust and waist. Hetty accepts Adam's proposal. Why, since she still loves Arthur, she should do so is not made immediately clear. But, at this stage, we may suspect a half-formed idea that if the wedding comes off quickly, she may still be able to pass off the child she fears she is carrying as Adam's—this, surely, is what a cunning dairymaid like Arabella Donn would have done in Hetty's place. (As will emerge later, I do not think this construction is entirely fair to Hetty.)

Assuming that Hetty was impregnated on her last encounter with Arthur, 18 August, and that date was midway between her periods, she will be, by 2 November, two months pregnant. Disastrously for her, Adam in his prudence postpones the wedding until March next, by which time he will have been able to build the extension to his mother's house, where they will live as man and wife. Over the subsequent months Hetty's forlorn hopes (that she is not after all pregnant, that she may be able to disguise the child as Adam's) slip away. By the time she runs away from Hall Farm to find Arthur at Windsor in February, Hetty must be six to seven months pregnant.

Why has no one, particularly the lynx-eyed Mrs Poyser, noticed Hetty's advanced condition? One reason is that, like Mrs Saddletree at a similar juncture in *The Heart of Midlothian*,[4] Mrs Poyser is made to be indisposed: 'confined to her room all through January' by a bad cold. But it beggars credulity that Mrs Poyser would not at some point over half-a-year have noticed the change in Hetty's shape, more so as they would have to share the same bathing facilities and the dairymaid would be obliged to strip off her outer clothing in the heat of her daily work. When Hetty faints in the Green Man public house the publican's wife, on unlacing the young woman, immediately perceives what is up (Ah, it's plain enough what sort of business it is!'). Why then is it not plain to Mrs Poyser? One may plausibly guess that Hetty's guardians *have* noted her condition. But they assume, logically enough, that Adam is the child's father. At this period of history in rural communities it was the rule rather than the exception that the bride would be with child at the altar. The Poysers might be surprised that the upright Adam would indulge in such immoral behaviour, but in a plighted couple premarital intercourse would not necessarily have been a matter of outrage or even much comment—so long as everything were made well in the church.

Hetty, of course, knows differently. In a passage in Chapter 35 recollecting her state of mind over the wretched months of her secret trial, there occurs the following: 'After the first on-coming of her great dread, some weeks after her betrothal to Adam, she had waited and waited, in the blind vague hope that something would happen to set her free from her terror.' Under its circumspect expression, this is one of the more remarkable passages in Victorian fiction. While the unworldly male and juvenile readers might construe the sense here as Hetty waiting for a letter or some miracle, no mature woman reader could miss the import of what is being said. The 'something' in early November that Hetty is desperately waiting for that will 'set her free from her terror' is her period.

What we may reconstruct is the following. After her last love-making with Arthur in mid-August, and the subsequent non-appearance of her period, Hetty was vaguely apprehensive (like him), but nothing more. As a demure young lady, brought up in a well-regulated religious household like the Poysers', Miss Sorrel would have had only the vaguest idea of where babies came from or—more precisely—the physiological signs and changes characteristic of early pregnancy. On her betrothal to Adam, it was appropriate (indeed a sacred duty) for Mrs Poyser to impart to her young ward 'the facts of life' (such maternal tuition was common well into the twentieth century, before schools got into the 'sex education' business). Now she knew what the missing period meant, Hetty experienced her 'great dread', and desperately prayed all through November and December that her period might belatedly come (Mrs Poyser would have told her that, for a girl her age, only recently past puberty, the occasional lapse of a month or so would not be unusual). When, as the year drew to its end, the 'something' did not happen, the full horror of her situation would have been tragically clear to Hetty: in her maidenly ignorance she had accepted Adam's proposal while carrying another man's child. Having told Hetty the facts of life, Mrs Poyser—noticing Hetty's interesting condition around January or February—would naturally assume that the young hussy could not wait to put her new knowledge into practice. Typical.

To return to the Showalters' claim with which this chapter began. This, surely, is a clear 'literary reflection' of the great unmentionable, menstruation. It is clearly something rather unusual, but doubtless there are other similar coded references in Victorian fiction that the radar of contemporary women readers might pick up.

The Oxford World's Classics *Adam Bede* is edited by Valentine Cunningham.

Wilkie Collins · *The Woman in White*

The missing fortnight

On its publication in three-volume form in August 1860 (after its triumphant nine-month serialization in *All the Year Round*) *The Woman in White* enjoyed a huge success, sparking off what today we would call a sales mania and a franchise boom. As Wilkie Collins's biographer Kenneth Robinson records:

> While the novel was still selling in its thousands, manufacturers were producing *The Woman in White* perfume, *The Woman in White* cloaks and bonnets, and the music shops displayed *The Woman in White* waltzes and quadrilles . . . Dickens was not alone in his enthusiasm. Thackeray sat up all night reading it. Edward FitzGerald read it three times, and named a herring-lugger he owned *Marian Halcombe*, 'after the brave girl in the story'. The Prince Consort admired it greatly and sent a copy to Baron Stockmar.[1]

Nuel Davis, in his life of Collins, goes so far as to claim that '*The Woman in White* was probably the most popular novel written in England during the nineteenth century'.[2] This is demonstrably untrue (*Robert Elsmere* and *Trilby* outsold Collins's novel by many times), but it is quite likely that it was the best-seller of the decade.

Among the chorus of applause there was one discordant voice. *The Woman in White* received a devastating review in *The Times* (then, as now, the country's newspaper of record) on 30 October 1860. In the review E. S. Dallas proved—by close scrutiny of dates in the crucial Blackwater Park episodes—that the events described in the novel *could never have happened*. *The Woman in*

White was, Dallas demonstrated, 'impossible'. As Dallas pointed out the whole of Collins's intricate denouement hinges on a single date—when was it that Laura took her fateful trip to London in late July 1850? If it can be proved that the date of Laura's journey from Blackwater to Waterloo station post-dated her recorded 'death' (in fact, the death of her look-alike half-sister, Anne Catherick) at 5 Forest Road, St John's Wood, then the criminals' conspiracy falls to the ground. With this crucial fact in mind, Dallas dismantled the plot machinery of *The Woman in White* with the ruthless precision of a prosecuting counsel exploding a shaky alibi:

> The question of a date is the pivot upon which the novel turns. The whole of the third volume is devoted to the ascertaining of this date. Everything depends upon it. But it is lost in the most marvellous obscurity—it is lost even to Mr Wilkie Collins, who is a whole fortnight out of his reckoning. If we dare trespass upon details after the author's solemn injunction [in his preface to the three-volume edition, that reviewers not give away the plot] we could easily show that Lady Glyde could not have left Blackwater Park before the 9th or 10th of August. Anybody who reads the story and who counts the days from the conclusion of Miss Halcombe's diary, can verify the calculation for himself. He will find that the London physician did not pay his visit till the 31st of July, that Dawson [the doctor who attends on Marian] was not dismissed till the 3rd of August, and that the servants were not dismissed till the following day. The significance of these dates will be clear to all who have read the story. They render the last volume a mockery, a delusion, and a snare; and all the incidents in it are not merely improbable—they are absolutely impossible.[3]

The details of Dallas's criticism are less important than its general thrust. What he was doing, and doing brilliantly, was subjecting a work of fiction to the criterion of falsifiability, in terms of its internal logic and structure. This test was something distinctly new in literary criticism, and a corollary of the fetishistic standards of documentary accuracy which Collins had imported into English fiction as his hallmark. As he says in his 'Preamble', Collins wanted his novel to be read as so many pieces of evidence, 'as the story of an offence against the laws is told in Court by more than one witness'. The reader, that is, should be as alert to clues and discrepancies in evidence as is a jury sitting in judgement. As Henry James astutely observed, Collins was playing a deep game with genre and literary discourse. Collins's novels, James declared, were 'not so much works of art as works of science. To read *The Woman in White* requires very much the same intellectual effort as to read Motley or Froude.'[4]

What James implied by comparing Collins to the leading historians of the age was that one could bring the same truth-tests to *The Woman in White* that a sceptical expert might bring to *The Rise of the Dutch Republic* (1855) or the *History of England from the Death of Cardinal Wolsey to the Defeat of the Spanish Armada* (12 vols., 1856–70). By reading *The Woman in White* as if it were history or science (rather than just a made-up story) Dallas can 'disprove' it.

Collins took Dallas's criticisms immensely seriously. He wrote to his publisher the next day, instructing that no more copies of *The Woman in White* must be put out, until he should have an opportunity to revise the text: 'The critic in "The Times" is (between ourselves) right about the mistake in time . . . we will set it right at the first opportunity', he confessed.[5] The mistake was duly set right in the 'New' 1861 one-volume edition by antedating the crucial Blackwater Park episode a whole sixteen days, and by clipping a couple of days off the crucial death-of-Anne/arrival-of-Laura episode (i.e. making it 25/26 July, rather than 28/29 July). In the revised edition, Collins made other small corrections (changing the wedding date of Percival Glyde and Laura, for instance, so that it did not hit a Sunday in the 1849 calendar—a document which the novelist evidently went back to consult in the course of revision).

Although he took extraordinary pains to reconcile fine points of narrative chronology, Collins left one troubling mote to trouble the reader's eye. In the revised 1861 text, a day or two after 20 June, when Marian falls into her fever, Count Fosco discovers Anne Catherick's whereabouts and treats her heart condition. Having won her confidence, the fat villain passes on to Anne a forged message from Laura, telling her to go to London with her old friend and companion, Mrs Clements. Laura, Anne is reassured, will meet her there. Three days later, Anne having been strengthened sufficiently by Fosco's medicines to undertake the journey, the two women leave for London, where they take lodgings. 'A little more than a fortnight' later (as Mrs Clements later testifies, p. 473) Anne is abducted. By Mrs Clements's reckoning, this must be around 7 July—two to three weeks before Anne's death. But, by Fosco's account in his final written confession to Hartright, it was on 24 July that Anne Catherick was abducted and brought to the house in St John's Wood as 'Lady Laura Glyde' (p. 623). The unfortunate woman died of heart failure there the next day, 25 July, and it was not until the day after that, 26 July, that the true Laura was lured to London. On 27 July she was returned to the London lunatic asylum as 'Anne Catherick', the true Anne Catherick now being prepared for her funeral at Limmeridge on 2 August, as 'Laura'.[6]

Who do we believe? Mrs Clements, by whose account Anne was in the sinister custody of the Foscos and the Rubelles for two weeks? Or Count Fosco, by whose account Anne was in their custody for two days? It does not require much inspection to see that Fosco's confession, for all its superficial candour, is shot through with self-serving falsehoods. Mrs Clements, by contrast, is stolidly honest. In terms of character the reader/jury will find her by far the more credible witness. And if Anne was held for two weeks in St John's Wood, what was she subjected to during that time? A clue is supplied in the famous anecdote of the novel's inspiration, given by J. G. Millais (the painter's son), ten years after Collins's death:

One night in the '50s [John Everett] Millais was returning home to 83, Gower Street from one of the many parties held under Mrs Collins's hospitable roof in Hanover Terrace, and, in accordance with the usual practice of the two brothers, Wilkie and Charles [Collins], they accompanied him on his homeward walk through the dimly-lit, and those days semi-rural, roads and lanes of North London . . . It was a beautiful moonlight night in the summer time and as the three friends walked along chatting gaily together, they were suddenly arrested by a piercing scream coming from the garden of a villa close at hand. It was evidently the cry of a woman in distress; and while pausing to consider what they should do, the iron gate leading to the garden was dashed open, and from it came the figure of a young and very beautiful woman dressed in flowing white robes that shone in the moonlight. She seemed to float rather than run in their direction, and, on coming up to the three men, she paused for a moment in an attitude of supplication and terror. Then, suddenly seeming to recollect herself, she suddenly moved on and vanished in the shadows cast upon the road. 'What a lovely woman!' was all Millais could say. 'I must see who she is, and what is the matter,' said Wilkie Collins, as, without a word he dashed off after her. His two companions waited in vain for his return, and next day, when they met again, he seemed indisposed to talk of his adventure. They gathered from him, however, that he had come up with the lovely fugitive and had heard from her own lips the history of her life and the cause of her sudden flight. She was a young lady of good birth and position, who had accidentally fallen into the hands of a man living in a villa in Regent's Park. There for many months he kept her prisoner under threats and mesmeric influence of so alarming a character that she dared not attempt to escape, until, in sheer desperation, she fled from the brute, who, with a poker in his hand, threatened to dash her brains out. Her subsequent history, interesting as it is, is not for these pages.[7]

Fosco, we remember, is a mesmerist: the Rubelles are thugs. What we can plausibly suppose is that, like the other luckless Woman in White, Anne was incarcerated for quite some time in a villa in the Regent's Park district, where she was subjected to barbarous mistreatment, which may well have included sexual abuse. It was this mistreatment which provoked her death from heart failure on 25 July. Fosco, out of guilt, suppresses the fact that he was responsible for Anne's death by torture, claiming instead that she died of 'natural causes', having been in his care only a few hours.

There is, of course, a simpler explanation—namely that Collins simply made another chronological miscalculation. But this seems unlikely. He revised the time-scheme of *The Woman in White* so conscientiously for the 1861 text, and he was so expert in such dovetailing, that it is much more attractive to assume that he left the anomaly for his more detectively inclined readers to turn up. This reading élite should have the privilege of knowing just how subtle and evil the Napoleon of Crime, Count Fosco, really was.

The Oxford World's Classics *The Woman in White* is edited by John Sutherland.

W. M. Thackeray · *Pendennis*
Mrs Gaskell · *A Dark Night's Work*
Anthony Trollope · *Rachel Ray*

Two-timing novelists

Early nineteenth-century novelists had an engagingly cavalier attitude to finer points of chronology. One of Dickens's footnotes in Chapter 2 of the 1847 reissue of *The Pickwick Papers* is typical of the freedoms they allowed themselves in such matters. Mr Jingle, in response to Pickwick's observation that 'My friend Mr Snodgrass has a strong poetic turn', replies: 'So have I . . . Epic poem—ten thousand lines—revolution of July—composed it on the spot.' When asked, he assures the amazed Pickwick that he was, indeed, there at the event and saw the blood flowing in the Parisian gutters. Dickens adds the footnote to this exchange: 'A remarkable instance of the prophetic force of Mr Jingle's imagination; this dialogue occurring in the year 1827, and the Revolution in 1830.'[1]

Scott is equally good-natured about his chronological solecisms in *Rob Roy*. When Andrew Fairservice urges the hero, Frank Osbaldistone, to accompany him to 'St Enoch's Kirk, where he said "a soul searching divine was to haud forth"', the novelist added the bland footnote (evidently as the result of a friend's observation): 'This I believe to be an anachronism, as St Enoch's Church was not built at the date of the story [1715].'[2] No more than Dickens, apparently, did Scott think of changing his anomalous text.

Scrupulosity about narrative chronology tightened up during the Victorian period, reaching a fetishistic pitch with the intricate sensation novels of Wilkie Collins. The older school of novelists were not, however, sure that they altogether liked the new orderliness about such things. Trollope voiced a typically bluff complaint in his comments on Collins in *An Autobiography:*

> Of Wilkie Collins it is impossible for a true critic not to speak with admiration, because he has excelled all his contemporaries in a certain most difficult branch of his art; but as it is a branch which I have not myself at all cultivated, it is not unnatural that his work should be very much lost upon me individually. When I sit down to write a novel I do not at all know, and I do not very much care, how it is to end. Wilkie Collins seems so to construct his that he not only, before writing, plans everything on, down to the minutest detail, from the beginning to the end; but then plots it all back again, to see that there is no piece of necessary dove-tailing which does not dove-tail with absolute accuracy. The construction is most minute and most wonderful. But I can never lose the taste of the construction. The author seems always to be warning me to remember that something happened at exactly half-past two o'clock on Tuesday morning . . . (pp. 256–7)

Trollope's underlying gripe would seem to be that by clock and calendar-watching Collins had deprived novelists and readers of valuable traditional liberties. In support of Trollope's preference for the old easygoing ways, we may examine chronological cruxes in three Victorian novels of the 1850s and 1860s, Thackeray's *Pendennis*, Mrs Gaskell's *A Dark Night's Work*, and Trollope's own *Rachel Ray*. Closely examined, they suggest that what looks like slovenliness about chronology in Victorian fiction can plausibly be seen as an artistic device which these three novelists, at least, used to powerful effect.

Pendennis was Thackeray's second major novel and it was, even by Victorian standards, an immensely long work (twenty-four thirty-two-page monthly numbers, compared to the twenty that made up *Vanity Fair)*. It took some twenty-six months in the publishing (November 1848–December 1850), interrupted as it was by the novelist's life-threatening illness which incapacitated him between September 1849 and January 1850. *Pendennis* is also long in the tracts of time its narrative covers. The central story extends over some forty years as the hero, Arthur Pendennis ('Pen'), grows from boyhood to mature manhood. In passing, Pen's story offers a panorama of the changing Regency, Georgian, Williamite, and Victorian ages.

Pendennis is one of the first and greatest mid-Victorian *Bildungsromanen*. The central character, as Thackeray candidly admitted, is based closely on himself. As part of this identification, Thackeray gave Arthur Pendennis the same birth-date as himself—1811. This is indicated by a number of unequivocal historical markers early in the text. Pen is 16 just before the Duke of York dies in 1827. We are told that Pen (still 16) and his mother recite to each other from Keble's *The Christian Year* (1827), 'a book which appeared about that time' (p. 31). Pen's early years at Grey Friars school and his parents' Devonshire house, Fairoaks, fit exactly with Thackeray's sojourns at Charterhouse school and his parents' Devonshire house, Larkbeare. Both young men go up to university ('Oxbridge', Cambridge) in 1829. Both retire from the university, in rusticated disgrace, in 1830. Thackeray clinches this historical setting by any number of references to fashions, slang, and student mores of the late 1820s and 1830s, as well as by a string of allusions to the imminent and historically overarching Reform Bill.

Switch from these early pages to the last numbers of the novel, where we are specifically told on a number of occasions that Pen is now 26. This is supported by any number of historical allusions fixing the front-of-stage date as the mid-to-late 1830s.[3] *Pendennis* is given its essentially nostalgic feel by historical and cultural events dredged up from the past, between ten and twenty years before the period of writing. And yet, there are a perplexing string of references which locate the action in the late 1840s, indeed, at the precise moment Thackeray was writing. In the highpoint scene of Derby Day in Number 19 of the serial narrative we glimpse among the crowd at the racecourse the prime minister, Lord John Russell, who took up office in 1846, and Richard Cobden MP. Cobden did not enter Parliament until 1847. And, by cross-reference to Richard Doyle's well-known Derby Day panorama in *Punch*, 26 May 1849 (a work which inspired Frith's famous *Derby Day* painting), we can see that Thackeray is clearly describing the Derby of the year in which he wrote, which actually took place a couple of months before the number was published.

Thackeray, as has been said, is insistent in this final phase of his narrative that Pen is just 26 years old (p. 797), which gives a historical setting of 1837. But, at the same time, Pen is given speeches like the following (one of his more provokingly 'cynical' effusions to Warrington), in Number 20:

> 'The truth, friend!' Arthur said imperturbably; 'where is the truth? Show it me. That is the question between us. I see it on both sides. I see it in the Conservative side of the House, and amongst the Radicals, and even on the ministerial benches. I see it in this man who worships by Act of Parliament, and is rewarded with a silk apron and five thousand a year; in that man, who, driven fatally by the remorseless logic of his creed, gives up everything, friends, fame, dearest ties, closest vanities, the respect of an army of Churchmen, the recognized position of a leader, and passes over, truth-impelled, to the enemy, in whose ranks he is ready to serve henceforth as a nameless private soldier:—I see the truth in that man, as I do in his brother, whose logic drives him to quite a different conclusion, and who, after having passed a life in vain

endeavours to reconcile an irreconcilable book, flings it at last down in despair, and declares, with tearful eyes, and hands up to heaven, his revolt and recantation. If the truth is with all these, why should I take side with any one of them?' (pp. 801–2)

No well-informed Victorian reading this in 1850 could fail to pick up the topical references. In his remark on the 'ministerial benches' Pen alludes to the great Conservative U-turn over the Corn Laws in 1846 (in which the aforementioned Cobden and Russell were leading players). His subsequent references are transparently to John Henry Newman (1801–90) and his brother Francis William Newman (1805–97). John went over to the Catholic Church in 1845. As Gordon Ray's biography records, Thackeray attended his course of lectures on Anglican difficulties at the Oratory, King William Street, in summer 1850 (this number was published in September). Francis, professor of Latin at University College London, 1846–69, published his reasons for being unable to accept traditional Christian arguments in the autobiographical *Phases of Faith* (1850). According to Gordon Ray, Thackeray was much moved by the book. Clearly Pen's comments to Warrington only make sense if we date them as being uttered in mid-1850, at the same period that this monthly number was being written. And we have to assume that Thackeray was making his hero the vehicle for what he (Thackeray) was thinking on the great current question of Papal Aggression.

These examples of 'two-timing' in *Pendennis* are systematic features of the novel's highly artful structure. Thackeray has devised a technique that was to be later explored and codified into a modernist style by the Cubists. Not to be fanciful, the author of *Pendennis* anticipates Picasso's multi-perspectival effect whereby, for example, more than one plane of a woman's face could be combined in a single image. Pen and his mentor

Warrington in the above scene are at the same time young men of the 1830s and bewhiskered, tobacco-reeking, 'muscular' hearties of the early 1850s, verging on middle-age. They inhabit the present and the past simultaneously, offering two planes of their lives to the reader.

Mrs Gaskell first published her novella *A Dark Night's Work* as a stopgap serial in Dickens's journal, *All the Year Round*. (Charles Reade needed more time to prepare for his massively documented sensation novel, *Hard Cash*.)[4] Mrs Gaskell's tale first appeared as instalments between 24 January and 21 March 1863 in the journal and was reissued as a one volume book later in the year. It would seem, however, that the work had an earlier origin. As the World's Classics editor, Suzanne Lewis, notes, 'although published in 1863, *A Dark Night's Work* was, according to a letter from Elizabeth Gaskell to George Smith, begun in about 1858'.[5] Lewis goes on to connect this double date (composition beginning in 1858, publication occurring in 1863) with an important dating reference in the story's first sentence: 'In the county town of a certain shire there lived (about forty years ago) one Mr Wilkins, a conveyancing attorney of considerable standing' (p. 1). The editor notes that by simple subtraction we may assume 'that the early events of the tale take place sometime between 1815 and 1820' (p. 304).

The earliest events told in the tale are Mr Wilkins's courtship of his wife which, thanks to a reference to the state visit of the allied sovereigns (p. 3), we can locate as taking place in the months directly following June 1814, when the Emperor of Russia and the King of Prussia were entertained in London. Mr Wilkins is subsequently widowed very young and left with his daughter, Ellinor, as his only consolation (a younger daughter dies a baby, soon after her mother). At 14, as we are told, Ellinor makes a friend of Ralph Corbet, who is studying with a clergyman-tutor nearby. And, four years or so later, the two young people are deeply in love. Meanwhile, Mr Wilkins has conceived a huge dislike for his obnoxiously efficient chief clerk, Mr Dunster, and his conveyancing business is going to the dogs.

Already some of the time-markers in *A Dark Night's Work* are beginning to go astray. As Suzanne Lewis notes, Ellinor, at some point shortly before 1829, is described visiting Salisbury Cathedral, where she earnestly discusses 'Ruskin's works' with the resident clergyman, Mr Livingstone (p. 47). Nothing of Ruskin's would have been available until decades later, and certainly not *The Seven Lamps of Architecture* (1849) and its extended discussions of the beauty of Salisbury Cathedral, which is what Mr Livingstone and Miss Wilkins are evidently supposed to be discussing.

Events crowd on quickly, driven by Mrs Gaskell's penchant for melodrama. In one of his drunken rages, Mr Wilkins strikes out and kills Dunster. In their 'dark night's work', Ellinor, the loyal servant Dixon, and Mr Wilkins bury the clerk's body, allowing it to be thought that he has stolen money from the firm and decamped to America. Under the strain of complicity, Ellinor falls ill; the engagement with Ralph Corbet is broken off; Mr Wilkins declines into chronic drunkenness and dies soon after, shattered by remorse and guilt. Ellinor is left virtually penniless, and goes to live a retired and self-punishingly religious life at East Chester. The narrative specifically dates the heroine's illness (and thus all the circumambient events) as 1829 (p. 116).

Ellinor remains at East Chester 'sixteen or seventeen years', before the long feared calamity occurs. Surveyors, prospecting for a new railway line between Hamley and Ashcombe, discover Dunster's body and an incriminating knife belonging to Dixon. Ellinor is at the time of the exhumation in Italy, and is unable to get back before the loyally mute Dixon is sentenced to death. The telegraph, we are informed, is not available to the heroine (as Suzanne Lewis notes, although introduced in the 1830s it did not come into general commercial service until the late 1850s: p. 312). Ellinor eventually does hurry back, by steamship and railway train, the bustle of which is graphically described.

The central chronological problem in *A Dark Night's Work* is explained by Suzanne Lewis, in a note to the '1829' reference to Ellinor's illness on page 116:

> This date does not fit the chronology of the story so far. When Ellinor is ill she is approximately nineteen; if the date of her illness is 1829, then the story begins before 1810, earlier than the date suggested by the 'forty years ago' of the opening paragraph, and the reference to the visit of the allied sovereigns in 1814. Moreover, a few days before she falls ill, Ellinor and Mr Livingstone discuss Ruskin's works, but even the earliest of Ruskin's works were not published until the mid 1830s. (p. 310)[6]

That Ellinor was, in fact, 19 at the time of the dark night's work is confirmed at the beginning of Chapter 12 where we are told 'her youth had gone in a single night, fifteen years ago' and the she is now 'only four-and-thirty' (p. 115). There would seem, therefore, to be a decade or more's slippage in the main events of the narrative, according to where in the text one is reading.

And then there is the strange business of characters' ages. After the murder, and the disposal of Dunster's body, while Ellinor is still recovering from her illness her accomplice, the lugubrious Dixon, is made to say to her:

> 'Ay! . . . We didn't think much of it at the time, did we, Miss Nelly? But it'll be the death on us, I'm thinking. It has aged me above a bit. All my fifty years afore were but as a forenoon of child's play to that night.' (p. 69)

By which we suppose that he was born in 1829 minus fifty years, that is, in 1779. But later on we are told that he solemnly takes Ellinor to the grave of his first love, 'Molly, the pretty scullery-maid'. The tablet over her grave reads: 'Sacred to the Memory of Mary Greaves, Born 1797. Died 1818. "We part to meet again"' (p. 125). 'I put this stone up over her with my first savings', Dixon tells Ellinor. Now clearly Dixon was not supposed to have been twenty years older than Molly, nor did he have to wait until he was 39 before he had sufficient savings to pay for a simple gravestone. Gaskell has created two Dixons, an old Dixon and a middle-aged Dixon—both of whom uneasily cohabit the narrative.

There is a similar dualism in the character of Miss Monro, Ellinor's governess, and another important secondary character in the tale. After Mr Wilkins's death and the news that there is no longer enough money in the household to employ her, Miss Monro rouses herself pluckily. She declares, 'I am but forty, I have a good fifteen years of work in me left yet' (p. 102). But, when we encountered her on her first arrival at Hamley, some ten years before, Miss Monro is described as 'a plain, intelligent, quiet woman of forty' (p. 11). Either she is gallantly lying when she tells Ellinor that she has fifteen years more of work in her or, more likely, we have another chronological anomaly: a character who does not age with the passing of time.

These anomalies in *A Dark Night's Work* seem to be genuinely irreconcilable. Nor, as with *Pendennis*, can one weave them into some artistic effect, aimed at by the author. They seem rather to belong to a broadbrush historical scheme of the 'before and after' kind in which the only thing that matters is whether something roughly succeeds or roughly precedes a threshold event. It would seem that Mrs Gaskell began *A Dark Night's Work* with a vivid *donnée*—the preparatory excavations for the railways churning up the beautiful English landscape, and discovering a murdered body. This *donnée* had social significance as well as directly personal implications for Ellinor. As Humphry House argues in *The Dickens World* (1941), the coming of the railways in the early years of the 1840s marked for the Victorians the passing of the old, 'innocent' England and the arrival of 'modernity', with all its woes. As a narrative (covering as it does some thirty years) *A Dark Night's Work* is divided into two halves, the world before the railways—a world of horses and sleepy country towns; and the world after the coming of the railways—marked in Gaskell's novel by joyless European tourism, frenetic rush, and the London where Ellinor has her fateful interview with Ralph Corbet.[7] The description of Ellinor's journey to Judge Corbet's house in Hyde Park Gardens to intercede for Dixon gives a vivid impression of the hectic pace of London life in the 1840s:

It was about the same time that she had reached Hellingford on the previous night, that she arrived at the Great Western station on this evening—past eight o'clock. On the way she had remembered and arranged many things: one important question she had omitted to ask Mr Johnson; but that was easily remedied. She had not inquired where she could find Judge Corbet; if she had, Mr Johnson could probably have given her his professional address. As it was, she asked for a Post-Office Directory at the hotel, and looked out for his private dwelling—128, Hyde Park Gardens. (pp. 154–5)

Ellinor sends a hotel messenger to the judge's house, but he is not at home, so she orders a cab for seven the next morning. All this hyperactive toing-and-froing belongs to the high-technology world of the 1840s and 1850s. In the 1820s and 1830s (when the first half of *A Dark Night's Work* is set) Ellinor would have had to lumber up to London by stage-coach and might have taken days tracking down Corbet by personal inquiry and trial and error. The Great Western Railway link referred to in the above passage was (as Suzanne Lewis notes) only completed in 1841. The Post Office Directory was, in the early 1840s, a recent innovation, which came in with Rowland Hill's reforms in the late 1830s. The fast and efficient metropolitan communication system (couriers and cabs arriving promptly at appointed hours) are all features of modern London, a London that has only recently come into being, we feel. In short, what Mrs Gaskell is aiming at in *A Dark Night's Work*, and what she succeeds in, is a novel divided historically into 'then' and 'now', with the arrival of the railway marking the point of division. It is a powerful effect, and one that lingers in the reader's mind.

Rachel Ray has one of the simplest of Trollope's plots. A young girl in an evangelically strict household receives a proposal of marriage from a young man. He is eligible in every way, except that he earns his money from brewing beer and has the reputation (ill-deserved) of being a 'wolf'. Torn between the moral anxieties of her widowed mother and widowed sister, Rachel neither accepts nor rejects Luke Rowan. She seems to have lost him by her vacillation, but finally he proves to be true.

Trollope chose to set *Rachel Ray* in Devon. The provincial setting has plausibly been ascribed to his admiration for George Eliot's recently published (and highly successful) *Scenes of Clerical Life* (1858). But Devon suggested itself for two other reasons. First, because it was historically a stronghold of evangelicalism (something that Trollope refers to in the novel on a number of occasions). Secondly, because Devon was a county Trollope knew intimately. Since childhood, he had spent holidays in Exeter with relatives. And in August 1851, on being transferred back to England from Ireland by the Post Office, he had been assigned to a roving mission to the West Country. 'I began in Devonshire', he records in his *Autobiography*, 'and visited, I think I may say, every nook in that county' (p. 88). Trollope loved Devon, and came to consider it the most beautiful region in England.

Rachel Ray vividly recalls Trollope's first impressions of Devon when he returned in the baking heat of summer 1851. He describes it with an uncharacteristically sensuous turn of phrase:

> in those southern parts of Devonshire the summer sun in July is very hot. There is no other part of England like it. The lanes are low and narrow, and not a breath of air stirs through them. The ground rises in hills on all sides, so that every spot is a sheltered nook. The rich red earth drinks in the heat and holds it, and no breezes come up from the southern torpid sea. Of all counties in England Devonshire is the fairest to the eye; but, having known it in its summer glory, I must confess that those southern regions are not fitted for much noonday summer walking. (pp. 16–17)

Peter Edwards, the editor of the World's Classics *Rachel Ray*, points out that for much of the manuscript Trollope wrote 'Kingsbridge' for 'Baslehurst' (where the Rays are supposed to live). The significance of the original name is that Kingsbridge is an actual town in South Hams, Devonshire. Trollope's diary records that Kingsbridge was virtually the first place that he visited when he came to the county, staying there on 9 August 1851. Coming as he had from cool, damp Ireland, he evidently remembered the heat of that summer week all his life.

One of the most distinctive features of Trollope—'the lesser Thackeray'—is that unlike his master, or Dickens, or George Eliot, he did not normally antedate the action of his fiction. His best-known novel for posterity is probably *The Way We Live Now* (serially published in 1874–5, set in 1873) and the title could describe most of the

Trollopian *oeuvre.* When he died in 1882, obituarists noted that if later generations wanted a 'photogravure' image of the Victorian age, they might consult Trollope's novels.

Rachel Ray would seem to deviate from the Trollopian norm in this respect. It is clear that in creating the novel's little world Trollope drew on his recollections of South Devon in August 1851. A number of cues in the action mark the action—or portions of it—as taking place well before 1863, the date of publication. So too does the generally sleepy atmosphere of Baslehurst. But at other points in the narrative Trollope specifically alludes to current events and even, at one point, seems to indicate a time-setting of August 1863 (which would be in line with his 'the way we live now' practice).

To summarize some of these chronological contradictions: in Chapter 4, we encounter the following, by way of introduction to the 'low' clergyman, Samuel Prong:

> As we shall have occasion to know Mr Prong it may be as well to explain here that he was not simply a curate to old Dr Harford, the rector of Baslehurst. He had a separate district of his own, which had been divided from the old parish, not exactly in accordance with the rector's good pleasure. Dr Harford had held the living for more than forty years; he had held it for nearly forty years before the division had been made, and he had thought that the parish should remain a parish entire,—more especially as the presentation to the new benefice was not conceded to him. Therefore Dr Harford did not love Mr Prong. (p. 51)

Edwards adds a note:

> the division of parishes such as Dr Harford's was carried out by the Ecclesiastical Commission, which was instituted in 1835 and was guided by a series of church-reform acts passed in the late 1830s and early 1840s. (pp. 407–8)

The 'Ecclesiastical Estates Commissioners' (set up, as Edwards notes, in 1835) had sweeping powers to equalize and rearrange dioceses, parishes, and clergymen's income. The re-parishioning referred to here must have taken place in the early 1840s at the latest, since the point is clearly made that Dr Harford has held the living for nearly forty years before the division and for just over forty years altogether. We must therefore assume the date of the narrative here also to be in the early to mid-1840s.

This date is confirmed by another reference, at the end of the same chapter, in the scene where Luke Rowan shows Rachel the beauty of the clouds at sunset. The narrator notes, sympathetically enough, that Luke 'was not altogether devoid of that Byronic weakness which was so much more prevalent among young men twenty years since than it is now' (pp. 54–5). The 'twenty years since' reference would also give a setting of the 1840s, which actually 'feels' right for this idyllic chapter.

There are, however, many other references in *Rachel Ray's* text which point to a present, 'way we live now' setting for the novel. There is, for example, an allusion to the early 1860s 'crinoline mania', in Chapter 7, in the description of the massed young ladies at the Tappitts' ball (p. 84).[8] In Chapter 9, as Dorothea contemplates the possibility of divorce, she clearly has the terms of the 1857 Matrimonial Causes Act, in mind (p. 123). There are passing references in the text to such modernistic things as night mail trains. And Mrs Ray's excursion by railway to Exeter in Chapter 21 clearly suggests a late 1850s, early-1860s date. Railways were relatively late to come to Devon (when he was posted there in 1851 Trollope was obliged to go about the county mainly by horse). The GWR did not offer a network service to a little town like 'Baslehurst' (i.e. Kingsbridge) until the early 1850s. In Chapter 21 ('Mrs Ray Goes to Exeter and Meets a Friend') Mrs Ray is described making a day trip to the nearby large town. Rachel commiserates with her mother on the older lady's return, observing she must be tired: 'Yes, I am tired, my dear; very' replies Mrs Ray; 'I thought the train never would have got to the Baslehurst station. It stopped at all the little stations, and really I think I could have walked as fast.' As the narrator tartly notes: 'A dozen years had not as yet gone by since the velocity of these trains had been so terrible to Mrs Ray that she had hardly dared to get into one of them!' (pp. 280–1). The dozen years would seem to refer the reader back to the early 1850s, when trains first arrived in the region. Finally, one may note that in the business of the election which dominates the second half of the narrative Mr Hart, the Jewish candidate, is clearly presenting himself after 1857–8, when it finally became possible for Jews to sit in Parliament (pp. 224, 412).

This two-timing in *Rachel Ray* does not give any impression of error, or narrative confusion. It creates a subtle and pleasing pictorial effect. Trollope paints an essentially contemporary Baslehurst—a thriving little country town of the 1860s, with its modern railways, breweries, and elections. But he throws over this depiction of rural modernity a kind of nostalgic halo, or veneer. Among all the contemporaneity we catch a momentary glimpse of old Devon, as it was fondly enshrined in Trollope's memory. For the reader of *Rachel Ray* it creates a peculiarly idyllic effect, like a sepia tint on a photograph.

In the three instances which have been examined here, structural anachronism, or the double time-setting, is not something one would necessarily want cleared up in any rationalistic spirit of standardized chronological reference. The case can be made, without excessive ingenuity, that these novelists' two-timing practices conduce to effects of historical complexity (in Thackeray's case), a highlighting of the major emotional divide in Victorian social history (in Gaskell's case), and a sensuously rich and cross-embroidered narrative texture (in Trollope's case). As Trollope insists in his *Autobiography*, too much scrupulosity in such matters may not always be the best thing artistically.

The Oxford World's Classics *Pendennis* is edited by John Sutherland; the Oxford World's Classics *A Dark Night's Work and Other Stories* is edited by Suzanne Lewis; the Oxford World's Classics *Rachel Ray* is edited by Peter Edwards; the Oxford World's Classics edition of Trollope's *An Autobiography* is also edited by Peter Edwards.

Anthony Trollope · *Phineas Finn*

The phantom pregnancy of Mary Flood Jones

Trollope has enjoyed three major revivals in the last fifty years. The first occurred during the years of Britain's siege culture in World War Two. Following a successful radio adaptation of the Barchester Novels in the early days of hostilities, reading Trollope became a cult among the literate middle classes as they whiled away the weary months of war. Dons in uniform like R. W. Chapman and John Sparrow, when they weren't cracking codes at Bletchley Park, exchanged erudite *aperçus* about textual minutiae in Trollope's novels. In a series of articles in the *TLS* they vied with each other to find misprints which had escaped the duller eyes of the novelist himself and his Victorian proof-readers.[1]

The second Trollopian wave came crashing in with the paperback classic reprints series which were pioneered by the Penguin English Library (now Penguin Classics) in the late 1960s. This series was followed by the revived Oxford University Press World's Classics, and the Everyman lines. There are, at the time of writing, up to four budget priced versions of most of Trollope's principal works of fiction in every high-street bookshop in the country.

The third and highest crest of the Trollopian wave came with the successful televisations of the Palliser novels in 1974. These six 'Parliamentary' novels, as Trollope called them (he studiously avoided the term 'political novels') were adapted for the small screen by Simon Raven. Raven was qualified for work on such a mammoth scale by virtue of his own ten-volume-strong 'Alms for Oblivion' saga (published between 1964 and 1975). He was used to the gigantic dimensions of the sequence novel. On the other hand, some viewers might well have thought Simon Raven disqualified by his raffish, and extremely un-Trollopian frankness about matters sexual in his 'Alms for Oblivion' narratives, in which fornication, adultery, and perversion were happenings as routine as breakfast.

In the event Raven did a magnificent job on 'The Pallisers' (as the BBC called the mini-series).[2] But as one might have expected (or feared) he imported into Trollope a vein of explicitness which was occasionally disturbing. This was notably the case with *Phineas Finn, the Irish Member*, the second novel in the sequence. The love plot of *Phineas Finn* is easily summarized. The young Irish hero (the son of a local doctor) wins the heart of a pretty Irish maiden, Mary Flood Jones, in his native Killaloe. He stands successfully for Parliament and leaves Killaloe for the flesh-pots of London. Before he goes, he snips a lock of hair from Mary's head, and steals a kiss. She duly regards herself as engaged. In London, Phineas MP falls in love with Lady Laura Standish, and proposes to her: she sensibly rejects the penniless young man (although in her heart she is drawn to him) in favour of a Croesus-rich Scottish landowner, Robert Kennedy. Phineas then turns his attention to a beautiful and spirited heiress, Violet Effingham. She in her turn rejects Phineas, finding a certain moral insubstantiality in a suitor who can direct his affections so briskly from one target to another. Finally, after three years and as a rising politician in dire need of wealth to support his ministerial career, Phineas allows himself to be proposed to by a brilliant society hostess, the immensely wealthy, widowed, and mysterious Madame Max Goesler. He tactfully declines, and returns to the faithful Mary. Almost penniless, they resolve to marry—on £150 a year, if necessary. Phineas is rescued from this awful fate by an appointment (at £1,000 p.a.) as a Poor Law Commissioner. His political friends have looked after him. The novel ends with a coy exchange between Phineas and Mary, indicative of his lover's impatience to enjoy his marital privileges:

'Oh Phineas; surely a thousand a-year will be very nice.'

'It will be certain,' said Phineas, 'and then we can be married tomorrow.'

'But I have been making up my mind to wait ever so long,' said Mary.

'Then your mind must be unmade', said Phineas. (ii. 356)

Phineas's reasons for giving up his London parliamentary career—which looks potentially glorious—are insufficiently explained by his good-natured difference of opinion with his Liberal leaders on a piece of Irish legislation. Nor is it entirely clear why he should reject Madame Goesler (whom he is, after trials and widowerhood, eventually to marry, in *Phineas Redux*).[3] Simon Raven solved this enigma by a daring stroke—he made Mary Flood Jones pregnant. Raven's Phineas, having weighed everything up, decided to do the right thing and make Mary an honest woman.

Although there is no textual warrant, there is a motivational logic in Raven's narrative improvement. It explains why Phineas—no fool, and ambitious to a fault—should have committed political suicide, as he did. Nor is it, as some commentators on the TV series complained, an 'Un-Trollopian' invention. Raven was no stranger to Trollope's fiction in 1974. He had earlier in 1966 adapted Trollope's great satire, *The Way We Live Now*, for television, and in the same year he had edited, for Anthony Blond's 'Doughty Library', Trollope's late novella, *An Eye for an Eye* (1879). In that story Fred Neville, a young English officer in the hussars, meets an innocent Irish beauty in Co. Clare. Kate O'Hara is cut from the same pattern as Mary Flood Jones. Fred (who is by no means a swine, but lacking in moral fibre) seduces and impregnates the luckless girl. Trollope does not (by the strict standards of Victorian fiction) mince his words:

> Alas, alas; there came a day in which the pricelessness of the girl he loved sank to nothing, vanished away, and was as a thing utterly lost, even in his eyes. The poor unfortunate one—to whom beauty had been given, and grace, and softness—and beyond all these and finer than these, innocence as unsullied as the whiteness of the plumage on the breast of a dove; but to whom, alas, had not been given a protector strong enough to protect her softness, or guardian wise enough to guard her innocence! To her he was godlike, noble, excellent, all but holy. He was the man whom Fortune, more than kind, had sent to her to be the joy of her existence, the fountain of her life, the strong staff for her weakness. Not to believe in him would the foulest treason! To lose him would be to die! To deny him would be to deny her God! She gave him all—and her pricelessness in his eyes was gone for ever. (pp. 108–9)

This echoes similar passages in *Phineas Finn*. In Chapter 16, for example, the hero returns to Killaloe for five months during the long recess of Parliament. He is now cool towards Mary, although he does not altogether discard her. Her mother warns her that Phineas intends to be false.

> Mary made no answer; but she went up into her room and swore before a figure of the Virgin that she would be true to Phineas for ever and ever, in spite of her mother, in spite of all the world—in spite, should it be necessary', even of himself. (i. 146)

In *An Eye for an Eye* Fred deserts a now-pregnant Kate, and returns to his grand life as future Earl of Scroope in England where he is expected to marry a young lady of his own class. He returns briefly to Ireland to put his affairs in order, and Kate's enraged mother pushes him over the cliffs to his death.

There is clear parallelism between the two plots. In both, a free-and easy young man wins a humble Irish girl's heart, and deserts her in order to make his way (and marry someone else) in England. He does not tell the women he pays court to in England about his forlorn Irish sweetheart. In one novel the deserted maiden loses a lock of hair, in the other her maidenhood. Raven was clearly justified in eliding details of the two plots, in the interest of stiffening his adaptation, and making it plausible to 1970s viewers.

The main objection to Raven's taking such a liberty is that it sets a up a train of subsequent contradictions to what Trollope manifestly intended. Mary Flood Jones (who is the most *disponible* of Trollopian heroines) dies in the un-narrated interval (two years, as we are told) between *Phineas Finn* and *Phineas Redux*. It is clear that Mary's death occurs very shortly before the action of the sequel (and thus at least a year after their marriage). As Laurence Fitzgibbon tells the party manager, Mr Ratler, 'the poor thing [Mary] died of her first baby before it was born' (p. 6). 'First baby' suggests that there was no little stranger on its way at the end of *Phineas Finn*. Trollope, one may guess, would have been mightily indignant about the slur cast on the honour of one of his more fragrant and virtuous young ladies.

There are, however, episodes elsewhere in the Palliser series which suggest that Trollope wants us to be aware that at every point more may be going on sexually than he is free to tell us about. There is, for instance, a revealing moment late in the action of the first of the Palliser novels, *Can You Forgive Her?* (1865). By this stage in the narrative the Byronic hero, George Vavasor, has gone utterly to the bad. He early on proposed to his cousin, the long-suffering

Alice, with the sole aim of getting her fortune. Since then his business speculations have gone wrong, he has been disinherited, and when his attempt to establish himself in Parliament fails, he assaults his faithful sister Kate. He compounds his falsehood by blackmailing Alice (who had given him her troth) and tries to murder his rival for her love, John Grey. At the depth of his misfortune, as he contemplates emigration or suicide, George receives a visitor. She reveals herself to be a discarded mistress, called Jane:

> She was a woman of about thirty years of age, dressed poorly, in old garments, but still with decency, and with some attempt at feminine prettiness. There were flowers in the bonnet on her head, though the bonnet had that unmistakable look of age which is quite as distressing to bonnets as it is to women, and the flowers themselves were battered and faded. She had long black ringlets on each cheek, hanging down much below her face, and brought forward so as to hide in some degree the hollowness of her jaws. Her eyes had a peculiar brightness, but now they left on those who looked at her cursorily no special impression as to their colour. They had been blue,—that dark violet blue, which is so rare, but is sometimes so lovely. (p. 321)

It emerges in the following conversation that for three years Jane was a kept woman, George's mistress. He installed her in a house. Then, having tired of her, he put her into a 'business' (a small shop, apparently), at a cost of £100. As he brutally puts it, 'That was all I could do for you;—and more than most men would have done, when all things are considered' (p. 323). The shop has failed, she is starving. If he cannot give her something to support herself, only one course is open to her—the streets. He dismisses her without a penny, at one point threatening to call the police if she does not go quietly. As we guess, she will be a whore before the week is out.

What is astonishing about this episode is that nowhere

earlier in the text is any sexual delinquency on George's part hinted at, even in the most distant or coded way. He is, albeit ruthless, a model of social propriety. He does not drink, gamble, or in any way waste his substance. His manners are perfect (even here he claims that he has acted better than other men would have done, 'when all things are considered'). And yet, we are to gather, all the time that we have had his acquaintance in the novel there was that secret compartment of his life about which we were not informed. The door swings open briefly—not to reveal anything very important, but merely to let us know that there are such doors and such secret chambers in Trollope's world.

What we may deduce from this scene in *Can You Forgive Her* is that, in the un-narrated hinterland of all of the Palliser novels, many of Trollope's gentleman characters have sexual lives which, just as they keep hidden from the world, the fiction keeps hidden from the reader. With this in mind, is it likely that a highly sexed, impulsive, handsome young buck like Phineas would keep himself entirely pure for five years in a foreign city? Raven's invention of a pregnant Mary Flood Jones is clearly recalcitrant in going so flatly against the grain of Trollope's narrative. On the other hand, Raven's blatant foregrounding of sex in *Phineas Finn* reminds us that Trollope's young heroes are neither neuters nor angels in their 'private', un-narrated lives.

The Oxford World's Classics *Phineas Finn* is edited by Jacques Berthoud. The Oxford World's Classics *An Eye for an Eye* is edited by John Sutherland. The Oxford World's Classics *Can You Forgive Her?* is edited by Andrew Swarbrick, with an introduction by Kate Flint.

George Eliot · *Middlemarch*

Is Will Ladislaw legitimate?

It helps to picture the *dramatis personae* of *Middlemarch* less as a community of English townspeople of the early nineteenth century than as a Papuan tribe—each connected to the other by complex ties of blood and marriage. Unknotting these ties requires the skills of the anthropologist rather than those of the literary critic. Let us start with Casaubon. Early in the narrative, the middle-aged vicar of Lowick is most vexed when Mr Brooke (making unwarranted deductions from their age difference) refers to young Will Ladislaw as 'your nephew'. Will, as Casaubon testily points out, is his 'second cousin', not his nephew. We learn from questions which Dorothea asks on her first visit to Lowick, that Casaubon's mother, whose Christian and maiden names we never know, had an elder sister, Julia. This aunt Julia—as we much later learn—ran away to marry a Polish patriot called Ladislaw and was disinherited by her family. Julia and her husband had one child, as best we can make out. Ladislaw Jr. (we never learn his first name) inherited from his father a musical gift which the son turned to use in the theatre—to little profit, apparently. In this capacity he met an actress, Sarah Dunkirk. Sarah, like Casaubon's aunt Julia, had run away from her family's household to go on the stage. At some point before or after running away, she discovered that her father, Archie Dunkirk, who at one point in the text is alleged to be Jewish (although a practising nonconformist Christian of the severest kind), was engaged in criminal activities. He is reported to have had a respectable pawnbroking business in Highbury, and another establishment which fenced stolen goods in the West End. Sarah broke off all relations with her mother and father on making this discovery. They have another child, a boy, and effectively disowned their disobedient daughter.

Sarah was subsequently married to, or set up house with, Ladislaw Jr. The couple had one child (Will) before the father prematurely succumbed to an unidentified wasting disease. Before dying, he introduced himself to Edward Casaubon, who generously undertook to take care of the penniless widow and child. Will is too young to remember anything distinctly about his father. On her part Mrs Ladislaw died in 1825, some ten years after her husband, of what is vaguely described as a 'fall'.

At some point after Sarah Dunkirk's breaking off all relations with her parents, her brother died. Shortly after this her father also died. In the distress of her double bereavement, Mrs Dunkirk (now an old lady, and—although she does not know it—a grandmother) turned to a young evangelical clerk of her husband's, called Nicholas Bulstrode. Eventually, she married the young man. The disparity in age precluded children. As she approached death, a distraught Mrs Bulstrode made desperate attempts to locate her daughter Sarah with the hope of reconciliation. But, although he had discovered their whereabouts, Bulstrode—assisted in his act of deception by another former employee of Dunkirk's, John Raffles—suppressed all information about Sarah and her little boy. Raffles in fact makes contact with Sarah twice, although he only informs Bulstrode about the first encounter (this is important since, for complicated reasons, he only discovers Ladislaw's name on the second occasion). When Mrs Bulstrode died, Bulstrode, by his act of deception, inherited his wife's entire fortune and used it to set up a bank in Middlemarch. It will help at this point to refer to a family tree (see Fig. 1). As Farebrother puts it with uncharacteristic coarseness later in the narrative (presumably echoing Middlemarch gossip): 'our mercurial Ladislaw has a queer genealogy! A high spirited young lady and a musical Polish patriot made a likely enough stock for him to spring from, but I should never have suspected a grafting of the Jew pawnbroker' (p. 588).[1]

This means, of course, that Bulstrode is Casaubon's distant cousin by marriage—although both are oblivious of the relationship. In Middlemarch, the still-young, newly widowed, and now rich Nicholas Bulstrode married Harriet Vincy. She is a sister of Walter Vincy, manufacturer, husband to Lucy Vincy and father of the novel's *jeunes premières* Fred and Rosamond. Mrs Vincy's sister was the second wife of the rich skinflint, Peter Featherstone. Featherstone's first wife was a sister of Caleb Garth, father of Mary Garth and other children who feature on the edge of the novel's plot. Peter Featherstone has no child by either of these two wives, both of whom predecease him by some years. But, unknown to his hopeful heirs (among whom the most hopeful is his nephew Fred Vincy), Peter Featherstone had a third, common-law wife, called Rigg. By this Miss Rigg, Featherstone has an illegitimate son, Joshua, born in 1798. Discarded by Featherstone, Miss Rigg subsequently married John Raffles, the aforementioned employee of Dunkirk and conspirator with Bulstrode to defraud Will Ladislaw of his inheritance. It will help here to refer to another tree (see Fig. 2).

Thus John Raffles, fence, blackmailer, gambler, the most despicable character in the novel, is related to Casaubon and (by subsequent marriages in the novel) to Lydgate, Dorothea Brooke, Will Ladislaw, and Sir James Chettam. The line of connection goes as follows: Raffles's wife is the mother of Peter Featherstone's child and heir; Peter

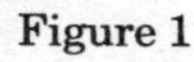

Figure 1

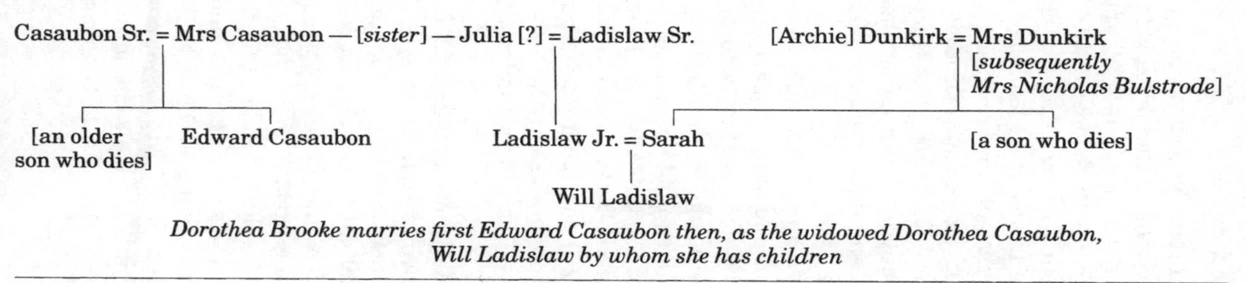

Dorothea Brooke marries first Edward Casaubon then, as the widowed Dorothea Casaubon, Will Ladislaw by whom she has children

---

Figure 2

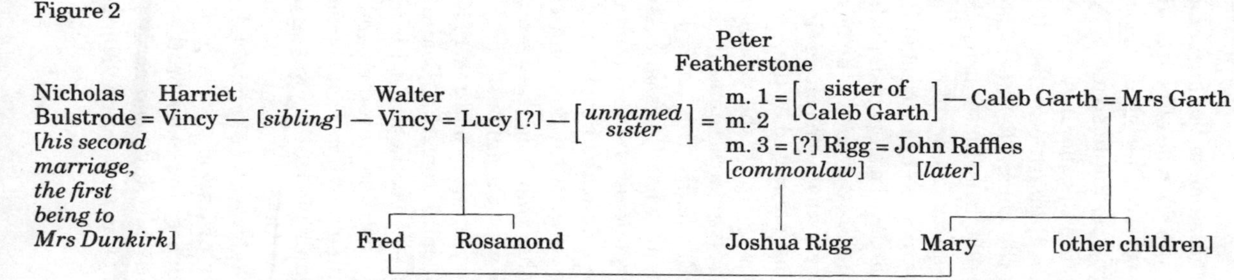

Fred Vincy subsequently marries Mary Garth, Rosamond Vincy marries Tertius Lydgate

Featherstone is the husband of Lucy Vincy's sister; Lucy Vincy is the wife of Walter Vincy, the mother of Rosamond (who marries Lydgate) and the sister of Harriet Bulstrode; Harriet is the husband of Nicholas Bulstrode; Bulstrode was the second husband of the former Mrs Dunkirk; Sarah Dunkirk was the wife of Ladislaw Jr.; Ladislaw Jr. was the cousin (by his aunt Julia) of Edward Casaubon; Casaubon is the husband of Dorothea; Dorothea is the sister-in-law of Sir James Chetham. Put simply, Bulstrode is (by marriage) Casaubon's cousin, Dorothea's cousin, Will's step-grandfather, and related in some way to virtually everyone in the novel.

Every main character in *Middlemarch's* massive plot can be connected by lines of consanguinity or marriage in this way—with the exception of Farebrother (who none the less regards himself as an 'uncle' to the Garth children). Clearly, it is part of Eliot's grand design, and pertains to what Rosemary Ashton aptly calls, 'the central metaphor of *Middlemarch*, the web'.[2] One of the odd features about the novel, however, is that we are not always sure how aware the main characters are of the webs of kinship that exist between them. This is particularly the case with Casaubon.

We know something of Casaubon's background from incidental remarks. Before their marriage, he tells his fiancée Dorothea that his mother was one of two daughters, just like Dorothea and Celia. Of his father we know nothing, except that Lowick is the family home and Casaubon's mother was a young woman there—whether she had been brought there as a bride, or had lived there as a child, we do not know. Edward was a younger son, and—after ordination—was given the living at Lowick. Subsequently his elder brother died (both parents were evidently already dead at this point) and he came into the manor as well as the vicarage. All these Casaubon deaths must have occurred much earlier, in the narrative's distant prehistory. It was evidently as the head of the family that Ladislaw Jr. approached Edward for help, and this was when Will was still too young to know anything about his circumstances other than that he was very hungry.

In this context one may note a remark which the Revd Cadwallader makes early in the narrative to Sir James Chettam. Sir James, still smarting at the absurd idea that Dorothea, whom he loves, should choose to marry such a dry stick as Casaubon, asks what kind of man he is. 'He is very good to his poor relations', Cadwallader says, using the plural form of the word.

> he . . . pensions several of the women, and is educating a young fellow [i.e. Will] at a good deal of expense. Casaubon acts up to his sense of justice. His mother's sister made a bad match—a Pole, I think—lost herself—at any rate was disowned by her family. If it had not been for that, Casaubon would not have had so much money by half. *I believe he went himself to find out his cousins*, and see what he could do for them. (pp. 56–7; my italics)

We never discover who these unnamed 'relations', 'women', 'cousins' are. They make no appearance at Casaubon's funeral, nor is any bequest to them mentioned in the subsequent lengthy discussion of the will. Dorothea inherits everything, as we understand (Will, who might have expected 'half', is spitefully excluded). What is interesting, however, is Cadwallader's recollection that Casaubon has made active investigations about his relatives, presumably on coming into his inheritance. He would surely have extended his inquiries to his aunt Julia and her offspring and might even have found out something about the murky world of the Dunkirks.

In conversation with Dorothea shortly after Casaubon's first heart-attack, Will casually tells her that his 'grandmother [was] disinherited because she made what they called a *mésalliance*, though there was nothing to be said against her husband, except that he was a Polish refugee who gave lessons for his bread.' Dorothea goes on to ask what he knew about his parents and grandparents, and Will replies:

> only that my grandfather was a patriot—a bright fellow—could speak many languages—musical—got his bread by teaching all sorts of things. They [i.e. grandfather and grandmother] both died rather early. And I never knew much of my father, beyond what my mother told me; but he inherited the musical talents. I remember his slow walk and his long thin hands; and one day remains with me when he was lying ill, and I was very hungry, and had only a little bit of bread. (p. 300)

Shortly after, Will recalls, 'my father . . . made himself known to Mr Casaubon and that was my last hungry day'. And shortly after that, his father died.

An initial mystery is why Julia's alliance with a cultivated Pole should have resulted in such a total alienation from her family. A second mystery is Casaubon's extraordinary disinclination to discuss anything to do with Will's origins. He returns all Dorothea's inquiries with 'cold vagueness' and refuses outright to answer any of his wife's questions about 'the mysterious "Aunt Julia"' (p. 305). Thirdly, one may wonder why Casaubon, an extraordinarily rectitudinous man, does not make over part of the family portion to Will, if he is the legitimate grandson of the older daughter (Julia) who would—had she not made her *mésalliance*—have inherited half or more of her parents' wealth, wealth which has all funnelled into the sole possession of Casaubon.

These questions, particularly the last, should be borne in mind when scrutinizing Raffles's account of how, all those years ago, he discovered Will and his mother and kept the news secret from Mrs Dunkirk, at Bulstrode's behest. 'Lord, you made a pretty thing out of me,' Raffles tells Bulstrode, on their reunion at Stone Court: 'and I got but little. I've often thought since, I might have done better by telling the old woman that I'd found her daughter and her grandchild: it would have suited my feelings better' (p. 431). He goes on, after revealing that Bulstrode gave him enough to emigrate comfortably to America,

> I did have another look after Sarah again, though I didn't tell you; I'd a tender conscience about that pretty young woman. I didn't find her, but I found out her husband's name, and I made a note of it. But hang it, I lost my pocket book. However, if I heard it, I should know it again . . . It began with L; it was almost all l's, I fancy. (p. 433)

The name, of course, is Ladislaw. This accounts why it is Bulstrode does not put two and two together when his young step-grandson turns up in Middlemarch. But under what name, then, did Raffles first discover the mother and child? One has to assume he discovered them as 'Sarah and Will Dunkirk', It beggars credulity that if he were charged to find proof of their identity, with £100,000 at stake, he would not have made some attempt to ascertain names and identities. Without some name to work with, how could he (or his lawyers) have found them in the first place?

This may be taken in conjunction with Ladislaw's extreme sensitivity on the subject of his mother. Consider, for example, Raffles's overture to Will, once he has tumbled to who he is. 'Excuse me, Mr Ladislaw', he asks:

> 'was your mother's name Sarah Dunkirk?'
>
> Will, starting to his feet, moved backward a step, frowning, and saying with some fierceness, 'Yes, sir, it was. And what is that to you?' . . . 'No offence, my good sir, no offence! I only remember your mother—knew her when she was a girl. But it is your father that you feature, sir. I had the pleasure of seeing your father too. Parents alive, Mr Ladislaw?'
>
> 'No!' thundered Will, in the same attitude as before. (pp. 497–8)

One notes that Raffles does not say—as would be normal—'was your mother's maiden name Sarah Dunkirk?' One notes too the extraordinary anger which Raffles's inquiry provokes. The same angry response is evident in Will's interview with the repentant Bulstrode, very shortly after:

> 'I am told that your mother's name was Sarah Dunkirk, and that she ran away from her friends to go on the stage. Also, that your father was at one time much emaciated by illness. May I ask if you can confirm these statements?'
>
> 'Yes, they are all true,' said Will . . . 'Do you know any particulars of your mother's family?' [Bulstrode] continued.
>
> 'No; she never liked to speak of them. She was a very generous, honourable woman,' said Will, almost angrily.
>
> 'I do not wish to allege anything against her . . . ' (pp. 507–8)

One notes again the use of the name 'Sarah Dunkirk' rather than, 'your mother's name was Sarah Dunkirk before marriage'. Also prominent is Ladislaw's touchiness at any aspersion against his mother's 'honour'.

There seems at least a prima-facie case for wondering whether or not Will was born out of wedlock. This would explain why it is he and his mother are first found by Raffles under some other name than Ladislaw (presumably 'Sarah Dunkirk and son'). Irregular unions were common enough in the nineteenth-century theatre world. It is worth recalling too that Will Ladislaw is known to be an idealized portrait of George Eliot's consort, G. H. Lewes. And, as Rosemary Ashton's recent biography has revealed, Lewes was illegitimate.[3] As Ashton notes, Lewes may not himself have known the fact. She adds: 'It is not surprising . . . that we know nothing about G. H. Lewes's earliest years. They must have been precarious socially, and probably financially as well . . . Whatever Lewes was told about his own father . . . he nowhere mentions [him] in his surviving writings' (p. 11). There is, of course, a difference. It seems that at some point Ladislaw Jr. did marry

Sarah Dunkirk—possibly as he felt death was coming, and he needed to hand over responsibility to Casaubon, he made an honest woman of Sarah and a legitimate child out of Will. All this is highly speculative. But any ideas we form about this aspect of the novel are driven to guess-work. One could, of course, speculate that Sarah when she left her parents took on a stage-name, and it was under this that Raffles first found her. There is also the baffling detail that at one point in the narrative Raffles seems to refer to Sarah's family name as 'Duncan' (see p. 498).

One is on firmer ground with hypotheses about what must have been going through Casaubon's mind, and agonizing him, as he watched Dorothea and Will forming a close relationship. He may well have known (from his interviews with Ladislaw Jr. and Will's mother, who has been alive until quite recently) about the discreditable 'Jew pawnbroker' business, if only vaguely. Should he tell the young man? Casaubon may also, as I have speculated, have known that Will was born out of wedlock (which would explain why he had not made part of the family fortune over to him, as a legitimate heir). At the very least, it seems strongly probable that Casaubon is possessed of some guilty knowledge and that the anxiety of it hastens his premature death.

The Oxford World's Classics *Middlemarch* is edited by David Carroll with an introduction by Felicia Bonaparte.

Anthony Trollope · *The Way We Live Now*

Is Melmotte Jewish?

In his monumental survey of anti-Semitism in the works of Thackeray, *Israel at Vanity Fair* (1992), S. S. Prawer refers in passing to Melmotte, the financier villain of Trollope's *The Way We Live Now*, as a 'Jewish crook' (p. 418). This is a common assumption among critics and readers of Trollope and can be found in print in any number of places. It is, very probably, a misapprehension—although it is not easy to come to a final resolution on the matter.[1]

The novelist apparently connives at our uncertainty about Melmotte's origins by not being as straightforward as he normally is on such matters. There is, for instance, no ambiguity about the race or religion of Samuel Cohenlupe, 'Member of Parliament for Staines, a gentleman of the Jewish persuasion' (i. 84), the amiable Ezekiel Brehgert, or the less amiable Mr Goldsheiner. Of Madame Melmotte, it is clearly explained not just that she is a Jewess, but that she is a particular variety of non-English Jewess—doubly alien: 'She was fat and fair,—unlike in colour to our traditional Jewesses; but she had the Jewish nose and the Jewish contraction of the eyes' (i. 31). Georgiana Longestaffe's accepting the proposal of Brehgert provokes an anathema worthy of a medieval Pope from her despairing, patrician, incurably xenophobic mother: 'It's worse than your wife's sister. I'm sure there's something in the Bible against it . . . An accursed race;—think of that, Georgiana;—expelled from Paradise' (ii. 263). 'Mamma, that's nonsense,' the hard-headed young lady responds, with the common sense of a commodity which has been twelve years' unbought on the marriage market and at the end of its shelf-life.

Although he is not generally mealy-mouthed about such things, Trollope deliberately, it seems, casts a pall of racial and national ambiguity around Melmotte. At our first introduction in Chapter 4 ('Madame Melmotte's Ball'), we are given an extensive but radically unclear description:

> The giver of the ball was Augustus Melmotte, Esq., the father of the girl whom Sir Felix Carbury desired to marry, and the husband of the lady who was said to have been a Bohemian Jewess. [In the manuscript of the novel 'Italian Jewess' is crossed out, and 'Bohemian Jewess' added.] It was thus that the gentleman chose to have himself designated, though within the last two years he had arrived in London from Paris, and had at first been known as M. Melmotte. But he had declared of himself that he had been born in England, and that he was an Englishman. He admitted that his wife was a foreigner,—an admission that was necessary as she spoke very little English. Melmotte himself spoke his 'native' language fluently, but with an accent which betrayed at least a long expatriation. Miss Melmotte,—who a very short time since had been known as Mademoiselle Marie,—spoke English well, but as a foreigner. In regard to her it was acknowledged that she had been born out of England,—some said in New York; but Madame Melmotte, who must have known, had declared that the great event had taken place in Paris. (i. 30)

Melmotte's speech, which should betray his origins, gives similarly mixed signals. When drunk, for instance, he speaks ungrammatically, and with what sound like Cockneyisms. Demanding to be introduced to the Emperor of China (for whom he has gone to the expense of organizing a lavish dinner) he declares: 'He don't dine there [i.e. in my house] unless I'm made acquainted with him before he comes. I mean what I say. I ain't going to entertain even an Emperor unless I'm good enough to be presented to him' (ii. 39). At other times, especially when addressing his wayward daughter Marie, Melmotte's native tongue—like his surname—seems clearly French. 'Pig! . . . wicked ungrateful pig!' (ii. 257), he shouts at his terrified and beaten daughter in the extremity of his rage. The objurgation does not sound like an English papa laying down the law to his errant offspring. But nowhere does Melmotte make errors in his speech, which is characteristically, in conversation with Englishmen, fluent and idiomatic. Nor is his speech, like that of Fisker, blemished with vulgar Yankeeisms—although, as we shall see, there are good grounds for thinking the Great Financier spent his formative years in New York.

Some clue to Melmotte's mysterious origins is given in an incidental reverie of Marie's, early in the narrative:

> She could just remember the dirty street in the German portion of New York in which she had been born and had lived for the first four years of her life, and could remember too the poor, hardly-treated woman who had been her mother. She could remember being at sea, and her sickness,—but could not quite remember whether that woman had been with her. Then she had run about the streets of Hamburg, and had sometimes been very hungry, sometimes in rags,—and she had a dim memory of some trouble into which her father had fallen, and that he was away from her for a time. She had up to the present splendid moment her own convictions about that absence, but she had never mentioned them to a human being. Then her father had married her present mother in Frankfurt. That she could remember distinctly, as also the rooms in which she was then taken to live, and the fact that she was told that from henceforth she was to be a Jewess. But there had soon come another change. They went from Frankfort to Paris, and there they were all Christians. (i. 106)

Although she was born in New York's 'German' quarter with English as the language of her country and city, the 19-year-old Marie has a poor grasp of English. She evidently spent some time at Hamburg, where her mother had died and Melmotte went to prison for fraud (see ii. 298). On his release, the widowed trickster evidently won himself a rich wife among the élite Jewish merchant class in Frankfort. Would Melmotte (or whatever he was then called)—a gentile—have been allowed to carry off a Jewish heiress and her dowry? And then there is the odd phrase in Marie's reverie, 'they were *all* Christians', with its implication that Melmotte converts, as well.

When he stands for Parliament as the member for Westminster, Melmotte attracts from his opponents and their newspaper allies the ironic title of 'the Great Financier'. This sobriquet is invented by the editor of the *Evening Pulpit*, Mr Alf. At this point in his career, despite his protestations (for electoral purposes) of invincible attachment to the Protestant faith (accompanied by some judicious charitable donations), 'it was suspected by many, and was now being whispered to the world at large, that Melmotte had been born a Jew' (ii. 52). To any moderately well-informed reader of 1875 it would have been crystal clear what Mr Alf's 'Great Financier' slur, and the pervasive rumours of his Jewish origins, alluded to. The Member of Parliament for the City of London at this time was Lionel Nathan de Rothschild. Rothschild was elected in 1847, but was not allowed to sit in the House because as a Jew he could not conscientiously take the oath. He was repeatedly re-elected as the most eminent financier in the City and finally took his seat in 1858. Rothschild was associated, like Melmotte (although more honourably), with lucrative French railway speculations. A dashing man, he was immortalized elsewhere in fiction as Sidonia, in Disraeli's *Coningsby.* (Trollope, incidentally, loathed Disraeli and all his works, both political and literary.) In 1872, Rothschild acquired the Tring Park estate, in Hertfordshire (as, in the novel, Melmotte schemes to acquire the Pickering estate), setting himself up as a landed

English nobleman. Trollope himself hunted with members of the Rothschild family in 1872, but seems to have fallen out with them, shortly before writing *The Way We Live Now.*

All this seems to point in a clear direction. But, in its last pages, the novel plays a last trick on us. After her father's death, Marie finds herself again an eligible heiress, with what little is left from the wreckage of Melmotte's ruin. She is claimed in marriage by the amiably amoral Hamilton K. Fisker. But even with Melmotte out of the way, there remain difficulties in establishing who Marie really is:

> At first things did not arrange themselves pleasantly between Madame Melmotte and Marie. The reader will perhaps remember that they were in no way connected by blood. Madame Melmotte was not Marie's mother, nor, in the eye of the law, could Marie claim Melmotte as her father. She was alone in the world, absolutely without a relation, not knowing even what had been her mother's name,—not even knowing what was her father's true name, as in the various biographies of the great man which were, as a matter of course, published within a fortnight of his death, various accounts were given as to his birth, parentage, and early history. The general opinion seemed to be that his father had been a noted coiner in New York,—an Irishman of the name of Melmody,—and, in one memoir, the probability of the descent was argued from Melmotte's skill in forgery. (ii. 449)

It is a plausible pedigree, and confirms details in Marie's faint childhood recollections, quoted above. But it remains tantalizingly vague, even at this last stage of the novel's complicated denouement.

So—is Melmotte English (as his command of the language and accent suggests), French (as his name, and occasional linguistic inflections suggest), American-Irish, or American-Jewish (i.e. sprung from the 'German' quarter of New York)? The clearest clue to what Trollope intended is to be found in his preliminary notes for the novel, which are held at the Bodleian Library, at Oxford. Putting together preliminary notes for his *dramatis personae*, Trollope jotted down the following thumbnail sketches:

~~Marianna Treegrene~~ Marie Melmotte, the heiress,
daughter of
~~Samuel~~ ~~Emanuel~~ Augustus ~~S. Treegrene~~ Melmotte
the ~~great American~~ French swindler
Madame Melmotte—fat Jewess.

It seems clear from these alterations that Melmotte began in Trollope's mind as two, and possibly three, separate conceptions. The first conception was Samuel Emanuel Treegrene, a Jewish swindler of German origins, who has anglicized his name from Grünbaum ('green tree'). The other conception was Augustus S. Melmotte, 'the great American Swindler', who has enriched himself by marriage to a rich Jewess in Frankfort, and by fraud in Paris, and who has Gallicized his name and Anglicized his pedigree the better to delude his victims and cover his criminal tracks. It would have been easy and logical for Trollope to have cleared up this confusion, by making Melmotte a thoroughly Jewish rogue (like Cohenlupe and Goldsheiner), or a thoroughly American rogue, like Hamilton K. Fisker. Instead, he dropped the tell-tale American initial ('Augustus *S.* Melmotte), and made his villain both, fusing the two conceptions and retaining a degree of irreconcileability, which he prudently shrouded by vagueness. Finally, he added to the mixture a dash of Irishness (drawing clearly on memories of the great Irish swindler John Sadleir, who committed suicide by cyanide—as does Melmotte—in 1856, and who was popularly presumed to be still alive in 1875, and living in New York). Melmotte, in short, is a national-racial compendium. As such, he can stand for a whole range of 'dishonesties' which the enraged Trollope perceived to be destroying the fabric of England, Europe, and America in the early 1870s (for the historical identity of the swindlers whom Trollope had in mind, see the Introduction to the World's Classics edition of *The Way We Live Now*).

As a satire against a whole gallery of foreign and domestic rascals and rascalities, Melmotte was aptly named. Trollope must surely have had in mind Charles Maturin's Gothic novel, *Melmoth the Wanderer* (1820). The hero, an Irishman, who is a version of the wandering Jew, cannot die for 120 years, and goes through many changes of identity in his long, largely villainous, career. Maturin's novel ends with a magnificent suicide by Melmoth, who casts himself off a crag at his ancestral estate in Wicklow (echoed in Melmotte's suicide in Trollope's novel). The theme of Maturin's novel was picked up by another 'cursed' Irish wanderer, Oscar Wilde, during his post-prison exile in Europe, where he took on the name 'Sebastian Melmoth'. The Christian-name alludes to the long-suffering saint, pierced by tormenting arrows but unable to die. 'Melmoth' echoes both Maturin (who was a distant relative of Wilde's) and Trollope's tormented villain.

Is Melmotte, to return to the question with which this chapter began 'a Jewish crook'? Yes, he is. But he is also a gentile, an Irish, an American, a German, and an English crook.

The Oxford World's Classics *The Way We Live Now* is edited by John Sutherland.

Anthony Trollope · *The Prime Minister*

Where is Tenway Junction?

The suicide of Ferdinand Lopez is regarded as one of the finest things in Trollope's late fiction. An adventurer of dubious foreign extraction (even he, we are told on the first page, does not know who, or of what nationality, or of what ethnicity were his grandparents), Lopez has successfully invaded the upper echelons of English society. By preying on the susceptibilities of Lady Palliser with his un-English smoothness of manner, he has almost gained election to Parliament. He has won a monied English bride. He has, for a while, been prosperous in the City of London. Finally, Lopez's house of cards falls. He is disgraced politically (bringing down Plantagenet Palliser's administration with him), wholly alienated from his wife (whom he has abused dreadfully), and ruined financially.

The last twelve hours of Lopez's life show him still going mechanically through the forms and rituals of a gentleman's existence. He takes dinner in his club in St James. He then undertakes a long peregrination on foot through nocturnal London. It is March, and a sleety, blustery night. He has an umbrella over his silk top hat. His route through the different quarters of the city is described by Trollope in meticulous topographic detail:

> he went round by Trafalgar Square, and along the Strand, and up some dirty streets by the small theatres, and so on to Holborn and by Bloomsbury Square up to Tottenham Court Road, then through some unused street into Portland Place, along the Marylebone Road, and back to Manchester Square by Baker Street. (ii. 188)

One can trace his circuitous route on any *London, A-to-Z*, and it seems that Lopez, as he stands on the brink of eternity, wishes to sees as many facets of fashionable, low-life, and bohemian London as is possible on a single night. On his way, he is 'spoken to frequently by unfortunates of both sexes' (the only reference that I know in decent Victorian fiction to male prostitutes, or 'rent boys', of the kind who were to bring Oscar Wilde to ruin in 1895). Once home in his Manchester Square mansion, Lopez comes face to face with the utter impossibility of his position. Early the next morning he leaves his house for the last time. It is still raining hard. He looks for a cab, but—as is invariably the case in London when it rains—none is to be found. In Baker Street he takes an omnibus (horse-drawn at this date) which carries him as far as 'the underground railway [i.e. what is now the Baker Street station], and by that he went to Gower Street [i.e. what is now the Euston Square station, on the Circle Line]' (ii. 190). As is his usual practice, Trollope has set the action of his novel very close to the present. The underground line (served by steam engines, and not a very attractive mode of transport) had been open since 1863. Lopez then crosses the Euston Road and walks 200 yards or so through the rain to the Euston Station. This great terminus, open since 1847, served in the 1870s as the gateway to the West and North. Again, Trollope gives us real names for these very real London places.

Lopez breakfasts on a mutton chop at the station café. Charmer to the end, he cannot forebear from flirting with the waitress. He then goes into the ticket-hall, and buys 'a first-class return ticket, not for Birmingham, but for the Tenway Junction'. The return ticket suggests that he has not yet decided irrevocably to kill himself—but there seems no other reason for him to go to Tenway Junction. Had he wished merely to disappear from the eyes of men under an alias, or abroad, he would surely have packed a bag and raised some cash. Trollope then embarks on a long description of Lopez's destination which opens, paradoxically, with the statement that any such description is, of course, wholly superfluous:

> It is quite unnecessary to describe the Tenway Junction, as everybody knows it. From this spot, some six or seven miles distant from London, lines diverge east, west, and north, north-east, and north-west, round the metropolis in every direction, and with direct communication with every other line in and out of London. It is a marvellous place, quite unintelligible to the uninitiated, and yet daily used by thousands who only know that when they get there, they are to do what some one tells them. The space occupied by the convergent rails seem to be sufficient for a large farm. And these rails always run into another with sloping points, and cross passages, and mysterious meandering sidings, till it seems to the thoughtful stranger to be impossible that the best trained engine should know its own line. Here and there and around there is ever a wilderness of wagons, some loaded, some empty, some smoking with close-packed oxen, and others furlongs in length black with coals, which look as though they had been stranded there by chance, and were never destined to get again into the right path of traffic. Not a minute passes without a train going here or there, some rushing by without noticing Tenway in the least, crashing through like flashes of substantial lightning, and others stopping, disgorging and taking up passengers by the hundreds. Men and women,—especially the men, for the women knowing their ignorance are generally willing to trust to the pundits of the place,—look doubtful, uneasy, and bewildered. But they all do get properly placed and unplaced, so that the spectator at last acknowledges that over all this apparent chaos there is presiding a great genius of order. (ii. 191–2)

Lopez, having contemplated the marvellous geometry of the Junction, walks down the bevelled end of a platform, into the path of the morning express from Inverness to London, which is coming round the curve into the station 'at a thousand miles an hour'. He is 'knocked into bloody atoms'. He who, we are told in the first chapter came from nowhere returns to oblivion.

The description of Tenway Junction and Lopez's suicide reveals an unexpectedly Dickensian aspect to Trollope's genius, recalling as it does Carker's symbolically charged death in *Dombey and Son*. There are added felicities: traditionally, suicides were buried in unhallowed ground at crossroads, and what more impressive crossroads has the world ever seen than this? There are, however, some baffling elements in the scene. As has been noted, Trollope opens with the declaration that it is unnecessary to describe something which he then proceeds to describe at great length. Baffling again is the assertion that 'It is quite unnecessary to describe the Tenway Junction, *as everybody knows it*.' Not every reader does. Elsewhere, to my chagrin, I have identified Tenway as Clapham Junction, which is clearly wrong, since Tenway is specifically indicated as being to the north of London and Clapham lies due south.[1] It must, presumably, be the Grand Junction at Willesden, in Middlesex: the point at which the (then privatized) Great Western, North London, London and North West, and Midland lines converged as they entered and left the Greater London network. (This identification is correctly made by the World's Classics editor, Jennifer Uglow.)

Most baffling, why does Trollope not give 'Tenway' its proper name? The pseudonym is not very brilliant (as Henry James pointed out, unlike his master, Thackeray, Trollope was not gifted in the 'science of names'). And pseudonymy jars with the accuracy of the earlier description of Lopez's night and morning peregrinations through London—all that cartographically exact business about Trafalgar Square, the Strand, and Baker Street. It may be, of course, that Trollope did not want to inspire copy-cat suicides. The Victorian was not as litigious an age as ours, but it may also be that Trollope did not want to libel the Willesden station-master by suggesting that his precautions against injury to passengers were insufficient.[2]

Plausible as these motives may be, it is more likely that Trollope fictionalized Willesden Junction for artistic reasons. Tenway Junction is not merely a busy railway station, any more than the great concourse which Thackeray describes in 'Before the Curtain' (in *Vanity Fair*) is merely a London street fair. Both are allegories of Victorian society. When he wrote *The Prime Minister* Trollope was well aware that his own death was not far off. He had given up the hunting he loved, and his health was failing. Like Lopez, he was standing on the edge of his own mortality. What Trollope desired at the climax of his narrative was a scene which should transcend the mere accidents of London topography. He wanted—like the hero of one of Browning's great monologues—to outreach himself, to exceed, for once, the Trollopian.

The Oxford World's Classics *The Prime Minister* is edited by Jennifer Uglow, with an introduction by John McCormick.

Anthony Trollope · *Is he Popenjoy?*

Was he Popenjoy?

Question-marks are everywhere in *Is he Popenjoy?*. They figure on the cover, at the head of chapters, and throughout the text. Questions are profusely interpolated into the authorial commentary and characters interrogate themselves and others constantly. There are, as I count them, some 1,175 question-marks speckling the narrative of *Is he Popenjoy?*. For what the statistic is worth, *Barchester Towers*, written twenty years earlier and much the same length, has only 626.

The big question, of course, is the titular one. In the novel as it opens, Mary Lovelace, the daughter of an amiably combative Dean (only a generation away from 'trade'), marries the younger son of an aristocratic family, Lord George Germain: Although he is not the heir, Lord George has very great expectations. His elder brother (by some ten years), the Marquis of Brotherton, lives in Italy, is in poor health, has dissolute habits, and is unmarried. Even if he does not himself inherit, Lord George may reasonably expect his son to do so. Heirs-apparent to the Marquisate are graced with the title 'Lord Popenjoy'. The Dean lives in eager expectation of having a Popenjoy as his grandchild.

George and Mary marry in June. An announcement is sent to the Marquis in Italy. He has 'never been a good correspondent' (i. 18), but the Marquis does on this occasion honour his brother with a communication of some three of four lines, indicating that (after his ten-year absence abroad) he may soon return to England, and that he will require George to vacate the grand family house. This is a blow; George is not rich, and he, his mother and sisters have grown used to occupying Manor Cross.

Then, 'about the middle of October', comes a bombshell in the form of a longer letter from Brotherton:

My Dear George,

I am to be married to the Marchesa Luigi. Her name is Catarina Luigi, and she is a widow. As to her age, you can ask her yourself when you see her, if you dare. I haven't dared. I suppose her to be ten years younger than myself I did not expect that it would be so, but she says now that she would like to live in England. Of course I've always meant to go back myself some day. I don't suppose we shall be there before May, but we must have the house got ready. (i. 51)

In addition to the distress of having to move into much less grand new accommodation, the letter also raises the possibility that the Marquis, with a ten-year-younger (i.e. in her thirties) wife, may well produce a Popenjoy.

Around November; at a dinner party in London, Mary hears more about the Marchesa, her new sister-in-law, from her neighbour at table, the middle-aged spinster, Miss Houghton:

'So the Marquis is coming,' [Miss Houghton] said. 'I knew the Marquis years ago . . . So he has married?'

'Yes; an Italian.'

'I did not think he would ever marry. It makes a difference to you—does it not?'

'I don't think of such things.'

'You will not like him, for he is the very opposite to Lord George . . . Have you heard about this Italian lady?'

'Only that she is an Italian lady.'

'He is about my age. If I remember rightly, there is hardly a month or two between us. She is three or four years older.'

'You knew her then?'

'I knew of her. I have been curious enough to inquire, which is, I dare say, more than anybody has done at Manor Cross.'

'And is she so old?'

'And a widow. They have been married, you know, over twelve months; nearly two years, I believe.'

'Surely not; we heard of it only since our own marriage.'

'Exactly; but the Marquis was always fond of a little mystery. It was the news of your marriage that made him hint at the possibility of such a thing; and he did not tell the fact till he had made up his mind to come home. I do not know that he has told all now.'

'What else is there?'

'He has a baby—a boy.' (i. 123–4)

Thus, through the malicious obliquities of social gossip, is the awful truth discovered. The Marquis (who told them in October that he was *going to be* married) has, it now emerges, actually been married for two years and has had an heir for twelve months. The lady is not, as he said in his letter, ten years younger than he, but a little older (i.e. well into her forties). At least there should be no more than the one child.

Lord George is appalled by the news—less for the confusion of his ambitions than at the bad manners of it all. The Marquis has not even told his mother, the dowager Marchioness, that she has a grandson. On his part the Dean is wildly suspicious. The mystery is compounded by reports in the newspapers, shortly before the Marquis's return with his new family. In one paper it is reported that the Marchesa Luigi, now the Marchioness of Brotherton, had been born in Orsini. In another paper it is reported 'that she had been divorced from her late husband' (i. 172). The Marquis's English agent is instructed, by telegram from Florence, to issue a public denial that the Marchioness is a divorced woman.

The Brothertons arrive at the end of April, a little sooner than planned. As they progress from the railway station to Manor Cross, 'the world of Brotherton saw them, and the world of Brotherton observed that the lady was very old and very ugly' (i. 199). The family are not permitted even to be introduced to the new Marchioness, nor view the baby. Dark suspicions gather: 'it was very odd that the marriage should have been concealed, and almost more than odd that an heir to the title should have been born without any announcement of such a birth' (i. 217). When pressed by a reluctant George on the matter, however, the Marquis airily observes that 'It's lucky that I have the certificated proof of the date of my marriage, isn't it?' (i. 218)—without actually producing this vital proof. When, belatedly, the young 'Popenjoy' is allowed to be seen by his English relatives, the child seems to Lord George 'to be nearly two years old. The child was carried in by the woman, but Lord George thought that he was big enough to have walked' (i. 228). If he is two years old, he must be either illegitimate or barely legitimate, assuming the marriage itself to have taken place two years earlier. But then, in his October letter, six months since, the Marquis clearly stated that he intended to get married, not that he had been already for some time a married man and a father. It is all very baffling.

The Dean, of course, has convinced himself that this 'Italian brat' who stands in the way of his daughter's ennoblement is illegitimate and that there has been criminal skulduggery. Lord George is prevailed on to write a letter, demanding from the Marquis some 'absolute evidence of the date of your marriage, of its legality, and of the birth of your son' (i. 251). The matter is put in the hands of a lawyer, the aptly named Mr Battle. Battle discovers that 'The Marchioness's late husband—for she doubtless is his lordship's wife—was a lunatic.' The plot thickens with the possibility of an earlier annulment. Battle continues, 'we do not quite know when he [the lunatic husband] died, but we believe it was about a month or two before the date at which his lordship wrote home to say that he was about to be married.' Then, asserts the Dean exultantly, the child cannot be legitimate! Not necessarily so, points out the more prudent lawyer: 'There may have been a divorce.' There is 'no such thing in Roman Catholic countries,' retorts the Dean, 'certainly not in Italy.' The lawyer is not sure, and adds 'I should not wonder if we found that there had been two marriages' (i. 278). The three men (Lord George, the Dean, Mr Battle) resolve to send a confidential clerk to Italy, to learn all the circumstances of what is evidently a very murky case.

The information which the lawyer's emissary brings back from Italy is perplexing. The Marquis, as suspected, had gone through two marriage ceremonies, 'one before the death, and one after the death, of her first reputed husband'. There was no divorce. But, 'Mr Battle was inclined, from all that he had learned, to believe that the Marchioness had never really been married at all to the man whose name she had first borne, and that the second marriage had been celebrated merely to save appearances' (ii. 35). Enquiries have been made 'at very great expense'. But no final determination can be made: 'the Luigi family [i.e. the family of the insane first partner] say that there was no marriage. Her family say that there was, but cannot prove it' (ii. 36).

Lord George is sufficiently convinced to give up the fight, although the Dean remains obdurate. The problem is finally solved by 'Popenjoy' dying. Contemplating his son's death (which he discovers affects him more than he expected), the Marquis remains bitter and venomous to the end. He confides to Mr De Baron, 'I'll go on living as long as I can keep body and soul together', adding:

> 'Poor little boy! . . . Upon my soul, I don't know whether he was legitimate or not, according to English fashions.' Mr De Baron stared. 'They had something to stand upon, but—damn it—they

went about it in such a dirty way! It don't matter now, you know, but you needn't repeat all this '

'Not a word,' said Mr De Baron, wondering why such a communication should have been made to him.

'And there was plenty of ground for a good fight. I hardly know whether she had been married or not. I never could quite find out.' Again Mr De Baron stared. 'It's all over now.'

'But if you were to have another son?'

'Oh! we're married now! There were two ceremonies.' (ii. 202)

This is the closest we ever come to a clear determination on the great question. Even the Marquis himself does not know whether his son was legitimate or not; or whether his wife is a bigamist. Nor did he ever know. Nor, apparently, will he ever know. Had the matter gone to trial, litigation would certainly have dragged on inconclusively for years. The question-mark attached to the title of the novel is as firmly attached at the end as it was at the beginning. Only the tense has changed. 'Was he Popenjoy?'—who can say?

Why did Trollope not close this gap, as he easily could have done? It is not his habit to leave such loose ends in his fiction. If ever a novelist liked tidy denouements, it was Anthony Trollope. The explanation is to be found, I suggest, in the dates of the novel's composition—12 October 1874 to 3 May 1875. The novel's title, as the Introduction to the World's Classics edition stresses, is highly topical. It alludes, as readers of the 1870s would immediately have picked up, to the absorbing question of the day: 'Is he Tichborne?' In 1854 Roger, the young heir to the Tichborne baronetcy and estates, had been drowned (as it was thought) at sea. Sir Roger's mother, a weak-minded lady, clung to the belief that her son was still alive. Her advertisements for information about her son were answered in 1865 by an apparent impostor, Arthur Orton, alias Thomas Castro—a loutish butcher from Wagga Wagga. With extraordinary effrontery, Castro came to

England to claim 'his' title and fortune. The family (except for the besotted mother) saw through Castro at once. He was brought to court first to eject him from his usurped title, then on charges of perjury. The case lasted from May 1871 to March 1874. In the way that some crimes do, 'Is he Tichborne?' caught the imagination of the British public. Orton won extraordinary sympathy among the English lower classes. There were public demonstrations and near-riots on his behalf in January 1874. Nevertheless—six months before Trollope began writing *Is he Popenjoy?*—the 'claimant' (as he was universally called) was found guilty, and sentenced to fourteen years' penal servitude.

Had the matter ended at this point, with Orton (Tichborne?) disposed of, Trollope might well have given the reader a clear answer to the 'Is he Popenjoy?' question in his ongoing novel. But the affair did not end in March 1874. In January 1875, with the opening of the new session of Parliament, the claimant's supporters introduced a slew of questions, notices, and petitions seeking a pardon for 'that unhappy nobleman now languishing in prison'.[1] They demanded a Royal Commission (a request that was contemptuously voted down by the House) and accused the judges who had passed sentence of being unfair to their man ('Sir Roger', as they obstinately maintained). Outside Parliament controversy also raged. On 29 March several thousand persons marched in procession with banners and music to Hyde Park for the purpose of making a demonstration in favour of the claimant.

All this put Trollope in something of a quandary. On his part, he had no doubts that the claimant was an impostor and his supporters fools and rogues.[2] But on 1 March he left England by boat for Australia. It was on board that he was obliged to write the last chapters of *Is he Popenjoy?*, finishing on 1 May, the day before landing at Melbourne. All this time he was effectively incommunicado, unable to get recent London papers. What might have happened in the meanwhile? Perhaps all the furore between January and March might have resulted in the verdict being overthrown? Perhaps Orton might have been shown by some legal *coup de théâtre* to be Sir Roger after all? Trollope, as his ship forged through the far Pacific, simply did not know how the 'Is he Tichborne?' question would be answered. Accordingly, the Popenjoy question is left similarly undecided.

The Oxford World's Classics *Is He Popenjoy?* is edited by John Sutherland.

Henry James · *The Portrait of a Lady*

R. H. Hutton's spoiling hand

The original text of *The Portrait of a Lady* has what is arguably the most delicately understated ending in all Victorian fiction. Isabel has at last come to a true appreciation of her husband, Osmond, and sees a straight path in front of her. In full consciousness of his moral worthlessness, and the worthiness of her loyal suitor, Caspar Goodwood, she has made a decision. The reader is not directly informed what that decision is. Osmond is meanwhile in Italy, and Isabel has been staying in London with her friend, Henrietta:

Two days afterwards Caspar Goodwood knocked at the door of the house in Wimpole Street in which Henrietta Stackpole occupied furnished lodgings. He had hardly removed his hand from the knocker when the door was opened and Miss Stackpole herself stood before him. She had on her hat and jacket; she was on the point of going out. 'Oh, good morning,' he said, 'I was in hopes I should find Mrs Osmond.'

Henrietta kept him waiting a moment for her reply; but there was a good deal of expression about Miss Stackpole even when she was silent. 'Pray what led you to suppose she was here?'

'I went down to Gardencourt this morning, and the servant told me she had come to London. He believed she was to come to you.'

Again Miss Stackpole held him—with an intention of perfect kindness—in suspense. 'She came here yesterday, and spent the night. But this morning she started for Rome.'

Caspar Goodwood was not looking at her; his eyes were fastened on the doorstep. 'Oh, she started—?' he stammered. And without finishing his phrase or looking up he stiffly averted himself. But he couldn't otherwise move.

Henrietta had come out, closing the door behind her, and now she put out her hand and grasped his arm. 'Look here, Mr Goodwood,' she said; 'just you wait!'

On which he looked up at her . . .

As originally published, the novel ends with Caspar's perplexed gaze into Henrietta's face—and very effective the ending is. Most readers assume that Isabel has of course gone back to Osmond. This is a willed, moral, unselfish, hugely courageous, 'ladylike' action. In her article 'Two Problems in *The Portrait of a Lady*', Dorothea Krook is contemptuous of any other construction which might be wished on to the end of James's novel by weak-minded readers:

> Why does Isabel go back to Osmond? This problem has, I believe, been somewhat artificially created for modern critics by a failure in critical perspective which arises from the disposition to ignore or minimize the context, historical and dramatic, in which Isabel Archer's final decision is made. I have heard it seriously argued that Isabel 'could after all have done something else'—walked out into freedom (like Nora in *A Doll's House*, presumably), or gone in for charitable works (like Dorothea Brooke in *Middlemarch*), or even perhaps taken a degree and become a pioneer in women's education, or whatever. The short answer to these bracing proposals is that Isabel Archer could have done none of these things. Her circumstances, historical, psychological, and dramatic—in particular the dramatic—absolutely prescribe any 'end' to her life other than marriage, and any duties, responsibilities or even serious interests other than those belonging to or arising out of that estate.[1]

Interestingly, Krook does not even consider; among the absurd alternatives open to Isabel, the absurdest of all: adultery or 'open sexual union' with Caspar. Yet this is one of the possible interpretations of Henrietta's 'Just you wait!'—that is, 'don't worry, she'll come back and then you two can get together at last'.

This may seem a far-fetched gloss to the last paragraphs of the 1881 *The Portrait of a Lady.* But this is precisely how the ending was misread in one of the most important reviews the novel received on its first appearance, that in the *Spectator* (unsigned, November 1881), by Richard Holt Hutton. 'Never before', Hutton thunderously declared in the peroration to his long piece, 'has Mr James closed a novel by setting up quite so cynical a sign-post into the abyss, as he sets up at the close of this book':

He ends his *Portrait of a Lady*, if we do not wholly misinterpret the rather covert, not to say almost cowardly, hints of his last page, by calmly indicating that this ideal lady of his, whose belief in purity has done so much to alienate her from her husband, in that it had made him smart under her contempt for his estimates of the world, saw a 'straight path' to a liaison with her rejected lover. And worse still, it is apparently intended that this is the course sanctioned both by her high-minded friend, Miss Stackpole, and by the dying cousin whose misfortune had been to endow her with wealth that proved fatal to her happiness.

Hutton delivered himself of much more in the same appalled vein, concluding with: 'We can hardly speak too highly of the skill and genius shown in many parts of *The Portrait of a Lady.* We can hardly speak too depreciatingly of the painting of that portrait itself, or of the moral collapse into which the original of the portrait is made to fall.'[2]

For most modern readers, the idea that Isabel is intending an eventual extra-marital liaison ('moral collapse') is grotesque. Hutton clearly does 'misinterpret' the last page in the most disastrous way. Yet, just as clearly, James worried about this misinterpretation. If an intelligent reader and former of public opinion like Hutton could go wrong, so might multitudes of others, When he revised the novel, together with all his major fiction, in 1908, he added the following postscript to the 1881 ending:

> . . . 'just you wait!'
>
> On which he looked up at her—but only to guess, from her face, with a revulsion, that she simply meant he was young. She stood shining at him with that cheap comfort, and it added, on the spot, thirty years to his life. She walked him away with her, however, as if she had given him now the key to patience.[3]

No one reading this version of the text could fall into Hutton's error. James here obliterates the remotest possibility that Isabel will ever throw in her lot with Goodwood. All that Henrietta means by her final instruction ('just you wait!') is that Caspar must be patient, and someone else will come into his life. He has many years ahead of him. It is hard not to feel that a fine Jamesian effect has been lost in order to achieve the corrective, unequivocal stress that the author felt was necessary.

Oddly enough, if Hutton were mounting the endings of major Victorian novels on his wall as trophies, *The Portrait of a Lady* would be joined by *Middlemarch*. The final two paragraphs of George Eliot's 'Study of Provincial Life' are as famous as anything she wrote. Never did the author achieve a more resounding tone of moral authority as—Godlike—she pronounced on the human condition in the middle years of the nineteenth century. None the less, Eliot clearly had authorial doubts about her concluding remarks. As David Carroll reveals in his World's Classics edition, not only did she have second thoughts while composing (see p. 707), she also made her most significant textual alterations in revised editions of *Middlemarch* to these last two paragraphs. Notably, she made a major excision in post-1874 versions of the novel (marked here by square brackets and italics):

> Certainly those determining acts of [Dorothea's] life were not ideally beautiful. They were the mixed result of a young and noble impulse struggling amidst the conditions of an imperfect social

state, in which great feelings will often take the aspect of error, and great faith the aspect of illusion [*They were the mixed result of young and noble impulse struggling under prosaic conditions. Among the many remarks passed on her mistakes, it was never said in the neighbourhood of Middlemarch that such mistakes could not have happened if the society into which she was born had not smiled on propositions of marriage from a sickly man to a girl less than half his own age—on modes of education which make a woman's knowledge another name for motley ignorance—on rules of conduct which are in flat contradiction with its own loudly-asserted beliefs. While this is the social air in which mortals begin to breathe, there will be collisions such as those in Dorothea's life, where in which great feelings will often take the aspect of error, and great faith the aspect of illusion.*] For there is no creature whose inward being is so strong that it is not greatly determined by what lies outside it. A new Theresa will hardly have the opportunity of reforming a conventual life, any more than a new Antigone will spend her heroic piety in daring all for the sake of a brother's burial: the medium in which their ardent deeds took shape is for ever gone. But we insignificant people with our daily words and acts are preparing the lives of many Dorotheas, some of which may present a far sadder sacrifice than that of the Dorothea whose story we know.

Her finely-touched spirit had still its fine issues, though they were not widely visible. Her full nature, like that river of which Cyrus broke the strength, spent itself in channels which had no great name on the earth. But the effect of her being on those around her was incalculably diffusive: for the growing good of the world is partly dependent on unhistoric acts; and that things are not so ill with you and me as they might have been, is half owing to the number who lived faithfully a hidden life, and rest in unvisited tombs. (pp. 682, 707–8)

As the editor of the Penguin Classics *Middlemarch*, Rosemary Ashton, records, Eliot made this substantial change 'after recognizing that those critics [notably Richard Holt Hutton] were right who pointed out that Middlemarch society did not smile on Mr Casaubon's proposal'. Hutton was thinking about such things as Mrs Cadwallader's sarcasms in Chapter 6 about the ill-assorted marriage.

Hutton had reviewed the serial parts *of Middlemarch* as they came out, in the *Spectator*, of which he was literary editor. He also reviewed the whole novel when it was published entire in December 1872.[4] He claimed to admire Eliot's achievement intensely. But he had doubts about the last paragraphs. Robert Tener and Malcolm Woodfield sum up Hutton's mixed feelings, and the seriousness with which Eliot took them:

> As Rosemary Ashton has shown, George Eliot's tone of irony towards her heroine disappears as the novel progresses until, in the Finale, Dorothea's disappointed life is blamed on 'the society which smiled on propositions of marriage from a sickly man to a girl less than half his own age.' Hutton was the first to point out the change of perspective (since, far from 'smiling' on the marriage, Dorothea's friends utterly disapproved of it). Following his objection, George Eliot paid Hutton the ultimate compliment, one which she afforded no other critic before or after, of cutting the passage from the 1874 book version, the version on which all subsequent editions have been based.[5]

As with *The Portrait of a Lady*'s Huttonian postscript, one may question whether the Huttonian excision from *Middlemarch* was artistically correct. It is true that Sir James Chettam and Mrs Cadwallader are scathing about an old stick like Casaubon (in his early forties) marrying a 19-year-old girl. Even Mr Brooke, in his ineffectual way, tries to dissuade his ward. But the fact is that Dorothea is two years short of the age of consent. Had Mr Brooke and her other guardians felt that strongly about the match, they could have forbidden it for twenty-four months with absolute authority and a clear conscience (Casaubon and Dorothea are the least likely people in all Victorian fiction to elope to Gretna Green). They (specifically Mr Brooke) consented to the match, even if 'smiling on it' may be slightly too strong a term. And why did they consent? Because Casaubon was a rich man. If he were a poor parson with just his living (that is, if his elder brother had not died, leaving him the wealthy heir of Lowick) there is no doubt but that Mr Brooke would have put his foot down very firmly. Eliot's essential point remains valid and could easily have been covered by toning down 'smiled on' to 'acquiesced in'.

There is another reason for regretting the excision. It is not just the one comment about sickly men marrying girls less than half their age which is removed. Eliot also took out the remark about 'modes of education which make a woman's knowledge another name for motley ignorance'. By removing this aggressive note of feminist protest, Eliot effectively depoliticized her ending. It represents a loss, since one of the sharpest themes in *Middlemarch* is that women's education must be improved if English society is to get the best from its women. Not to make this point explicitly and with force is to muffle a conclusion which the novelist has worked hard for.

Finally, there is the rather mysterious remark which Eliot modified about great feelings 'colliding' with society's 'loudly-asserted beliefs' about 'rules of conduct', thus presenting to crasser minds the aspect of error. I would guess that what Eliot is thinking about here is her own unsanctified union with G. H. Lewes. Technically, she and Lewes were living in sin. This is what small-minded contemporaries said about them and, as Gordon Haight records in his biography, 'Mr and Mrs Lewes' (as they liked to be known) encountered persistent social ostracism for their having sacrificed conventional decencies to their 'great feelings'.[6]

In general, Hutton's approach to narrative is one which will be congenial to many readers. He was a close and often bloody-minded scrutinizer of the text, forever looking for nooks and crannies into which to introduce a pedantic or a downright perverse interpretation. But I would suspect that George Eliot was not driven to make her massive (for such it is) change to the conclusion of *Middlemarch* merely because Hutton had perceived a slight contradiction between what was said in the penultimate paragraph and what had been said by some second-rank characters in Chapter 6. What alarmed her was the fact that someone as authoritative as the literary editor of the *Spectator* was looking so closely at the last sentences of her novel. So as not to confuse the moral peroration with secondary distractions, she purged her conclusion not just of its small narrative anomaly (which did not in fact require anything more than a change of phrase) but of its sexual politics and a significant measure of its personal import.

Both Eliot and James made the claim, common enough with great novelists, that the opinions of the critics meant nothing to them. But it is unarguable that in the above instances the reviewer's opinions were not just registered, they were deferred to. Of the millions of twentieth-century readers who have read *Middlemarch* in its post-1874 version and *The Portrait of a Lady* in its post-1908 version, probably less than one in ten thousand could say who Richard Holt Hutton was. And yet, for all time, his messy fingerprints will be all over the conclusions to these novels.

The Oxford World's Classics *The Portrait of a Lady* is edited by Nicola Bradbury.

R. L. Stevenson · *Dr Jekyll and Mr Hyde*

What does Edward Hyde look like?

Say 'Jekyll and Hyde', and the person you are speaking to will see in the mind's eye Spencer Tracy's amiably pudgy features dissolving into the monstrous physiognomy of Edward Hyde. The transformation is one of the high-points of early Hollywood special effects and close-up camera artistry. The World's Classics volume pays tribute to the 1941 film by carrying on its cover a still photograph of Tracy in his Hyde make-up.[1] Thanks to the film, when they think of Jekyll and Hyde modern readers see a face very vividly and in great detail. One of the puzzles in Stevenson's source text, however, is that where Hyde's face ought to be in the narrative there is a blank—rather like the facial features technologically fuzzed out in television documentaries and newscasts 'to protect the innocent'. What precisely was Stevenson's motive for doing this?

The numerous descriptions of Edward Hyde in the narrative agree on a number of points: (1) he is physically small; (2) he is a 'gentleman'; (3) he is young; (4) he is, in some unspecified way, 'deformed'. Jekyll, by contrast, is 'a large, well-made, smooth-faced man of fifty', possessed of a 'large handsome face' (pp. 22–3). The first eyewitness description which we are given of Hyde arises out of the 3 a.m. outrage when he tramples a little girl in the London street (only children and the very aged are at risk from Edward Hyde, who is not, it seems, inclined to tackle fully grown, physically active adults). According to Mr Enfield, the assailant is 'a little man'. That he is also a 'gentleman' is testified to by his having a bank account at

Coutts and his evening dress. Enfield gives no description of the monster's face, but recalls that it was his manner, his 'black sneering coolness', which infuriated the onlookers. Describing the nasty episode to Utterson, Enfield strikes what is to be a recurrent note in response to the question 'What sort of man is he to see?' He is not easy to describe, Enfield recalls:

> There is something wrong with his appearance; something displeasing, something downright detestable. I never saw a man I so disliked, and yet I scarce know why. He must be deformed somewhere; he gives a strong feeling of deformity, although I couldn't specify the point. He's an extraordinary looking man, and yet I really can name nothing out of the way. No, sir; I can make no hand of it; I can't describe him. And it's not want of memory; for I declare I can see him at this moment. (p. 12)

Playing the part of 'Mr Seek', Utterson himself comes across Mr Hyde in the streets after one of his horrible nocturnal adventures. The encounter produces the following Identikit picture: 'He was small and very plainly dressed, and the look of him . . . went somehow strongly against the watcher's inclination . . . Mr Hyde was pale and dwarfish, he gave an impression of deformity without any nameable malformation' (pp. 17–19). Later, arising out of the Carew murder case, the anonymous maidservant who witnessed the crime describes 'a very small gentleman'. This is the best she can do, although she manages a fuller description of Sir Danvers Carew, MP, 'an aged and beautiful gentleman with white hair'. What colour, then, was Hyde's hair? Evidently her eyes are sharp enough to pick up such detail, but something prevents her 'seeing' the assailant. Her identification is exclusively moral: the murderer was 'particularly small and particularly wicked looking' (pp. 25–7). The chapter ends with another highly charged passage which expatiates on Hyde's inscrutability, his powerful nonentity:

He had numbered few familiars . . . his family could nowhere be traced; he had never been photographed; and the few who could describe him differed widely, as common observers will. Only on one point, were they agreed; and that was the haunting sense of unexpressed deformity with which the fugitive impressed his beholders. (p. 29)

It is worth pausing here because, over the last fifty years, Mr Hyde has been very widely 'photographed'—that is to say, he has been depicted in innumerable films and television adaptations. These versions of Mr Hyde invariably agree on what he looks like: simian, excessively hairy, thick-lipped, beetle-browed, swarthy, middle-aged, and physically massive (see the World's Classics cover). One of the 'artist's impressions' loved by newspapers would be very easy to draw up. None the less, in Stevenson's source narrative Hyde continues through the middle and later stages shrouded in physical and physiognomic vagueness. There is, for instance, the strange episode in which Jekyll's servant sees Edward Hyde in his master's laboratory and assumes that—like the Elephant Man—he wears 'a mask upon his face' (p. 45). His subsequent description is as unhelpful as those of all the other witnesses have been ('When that masked thing like a monkey jumped from among the chemicals and whipped into the cabinet, it went down my spine like ice'). When, in the climax of the melodrama, Dr Lanyon confronts Edward Hyde face to face, any clear description of the monster is once again withheld:

Here, at last, I had a chance of clearly seeing him. I [Lanyon is writing] had never set eyes on him before, so much was certain. He was small, as I have said; I was struck besides with the shocking expression of his face, with his remarkable combination of great muscular activity and great apparent debility of constitution, and—last but not least—with the odd, subjective disturbance caused by his neighbourhood. (p. 56)

On his part, looking at the twitching dying visage of Hyde, Utterson declares, 'the cords of his face still moved with a semblance of life, but life was quite gone' (p. 49). All this tells us is that Hyde is possessed of a face. What that face's lineaments or features are we can only guess.

The sum of the physiognomic description which we are given is very unsatisfactory. The only concrete detail we have is that Hyde's face is 'pale', which does not much help the mind's eye picture him. Henry Jekyll's own final testament adds one small detail, which has been eagerly seized on by film-makers. It occurs during an unwilled metamorphosis when, to his horror, Dr Jekyll discovers he has changed personality during a 'comfortable morning doze':

> I was still so engaged when, in one of my more wakeful moments, my eye fell upon my hand. Now the hand of Henry Jekyll (as you have often remarked) was professional in shape and size: it was large, firm, white and comely. But the hand which I now saw, clearly enough, in the yellow light of a mid-London morning, lying half shut on the bed clothes, was lean, corded, knuckly, of a dusky pallor and thickly shaded with a swart growth of hair. It was the hand of Edward Hyde. (pp. 66–7)

Taking this cue, special-effects departments have gone straight for the box marked 'Werewolf'. In the famous Lon Chaney movie, the hero is seen by the technique of stop-frame photography transmuting in front of the camera's eye from respectable middle-class man to disgusting hairy monster. So, too, does Spencer Tracy transmute in the film, *Jekyll and Hyde.* In fact, there is no warrant for assuming a hairy, monstrous Hyde. This is the only reference to hairiness which we are given. There are many more countervailing references in the text to his being 'pale' and 'childlike'; It may well be that a distraught Jekyll is not seeing himself clearly at this point—he may be hallucinating. It is more likely that Hyde now looks different from what he initially did. Jekyll goes on to say in his 'Statement of the Case':

> That part of me which I had the power of projecting, had lately been much exercised and nourished; it had seemed to me of late as though the body of Edward Hyde had grown in stature, as though (when I wore that form) I were conscious of a more generous tide of blood . . . (p. 67)

Following the logic of this statement, Mr Hyde should be conceived as changing from young, hairless, juvenile monster in his early manifestations to the hirsute, middle-aged, bulkier monster of his last appearance. No such distinction is traditionally made in film versions. To summarize: there are, it seems, some very good photographs of Mr Hyde—but they are not Stevenson's. In his album there is a perfect blank, apart from the tantalizing snapshot of one enigmatic hand.

The Oxford World's Classics *Dr Jekyll and Mr Hyde* is edited by Emma Letley.

R. L. Stevenson · *The Master of Ballantrae*

Who is Alexander's father?

The main element in Robert Louis Stevenson's mind as he embarked on *The Master of Ballantrae* was the idea of an old Scottish family hedging its bets at the time of the 1745 Revolution by dispatching one son to serve with the Pretender while another son remained at home, a loyal servant of King George. Stevenson had an actual historical case in mind.[1] It is, in the story, quite arbitrary which brother will join the rebels. The question is decided by the flip of coin, as are other crucial decisions in the Durie brothers' careers. But what began as an eminently sensible hedging of bets returns as a curse which haunts and eventually destroys the House of Durrisdeer and Ballantrae.

An early, discarded title for the new novel was 'Brothers'. Another was 'The Familiar Incubus'. The incubus in *The Master of Ballantrae* is self-evidently James—the elder brother (or 'Master') who is 'killed' first at Culloden, then in the duel with his brother, and finally in the frozen Adirondacks. On each occasion he rises from the grave to haunt his brother Henry ('Jacob', as he tauntingly calls him, for having stolen the birthright of Esau). The Master, James, finally dies only after he has succeeded in frightening his luckless victim into dying first. It is a surreal conception. And one can perhaps track its meaning by looking more closely into the meaning of 'incubus'. Its primary sense is, as the *OED* tells us: 'A feigned evil spirit or demon, supposed to descend upon persons in their sleep, and especially to seek carnal intercourse with women.' Sexual predatoriness is the salient characteristic of the incubus. One of the most masterly elements in the telling of this most obliquely narrated of tales is the way in which RLS manœuvres the reader into calculations and reckonings which 'prove' adultery, illegitimacy and incest. These things are never actually made clear, but the reader, almost unwillingly, cannot but be conscious of them.

The narrative of *The Master of Ballantrae* opens with the protestation by Ephraim Mackellar that he will give the world 'The full truth of this odd matter'. A couple of pages later, it is made clear that where the 'full truth' is concerned, Mr Mackellar can sometimes be less than forthcoming. Of the early scapegrace history of James (the elder, and from childhood the more wayward of the brothers) he writes:

> One very black mark he had to his name; but the matter was hushed up at the time, and so defaced by legends before I came into those parts that I scruple to set it down. If it was true, it was a horrid fact in one so young; and if false, it was a horrid calumny. (p. 11)

What this 'black mark' is, we never learn. It must be something more than merely spawning bastards and spurning their mothers (Ephraim is later forthcoming enough on that score). Something of the order of rape or murder seems to be alluded to.

The Durie brothers are brought up in the same household with Alison Graeme, 'a near kinswoman, an orphan, and the heir to a considerable fortune'. She and James fall in love, and—we may assume—possibly their love has been consummated. The Master is not one to rein his appetites back in such matters. After the flip of the coin, which decides that James shall join the rebels, the following exchange occurs between the Master and his young mistress:

'If you loved me as well as I love you, you would have stayed', cried she.

'"I could not love you, dear, so well, loved I not honour more",' sang the Master.

'O!' she cried, 'you have no heart—I hope you may be killed!' and she ran from the room, and in tears, to her own chamber. (p. 13)

James goes off to fight with the Pretender, and is reported to have been killed at Culloden. In her grief, Alison upbraids Henry: 'There is none but me to know one thing—that you were a traitor to him in your heart' (p. 16). The significance of this remark is obscure, like much of the sexual underplot of this story.

Although he inherits the estate, Henry is much maligned. Survivors of the Jacobite faction in the neighbourhood call him 'Judas'. As Ephraim recalls: 'One trollop, who had had a child to the Master [i.e. James], and by all accounts very badly used, yet made herself a kind of champion of his memory. She flung a stone one day at Mr Henry' (p.18). The trollop, Jessie Broun, witnesses to the sexual ruthlessness of James, and his none the less irresistible attraction to the women he preys on.

'Miss Alison's money was highly needful for the estates' (p. 16), Ephraim, the steward, notes. Marriage will ensure the survival of the house. In what may be gratitude for the shelter she has received over the years, Alison finally accepts Henry's proposal with the bleak statement: 'I bring you no love, Henry; but God knows, all the pity in the world' (p. 20). They marry, as Ephraim precisely records, on 1 June 1748. By the end of December, as he again precisely records, Alison's baby is 'due in about six weeks' (p. 22). The conjunction of dates is a clear instruction to the reader to do some reckoning. If the marriage took place on 1 June and the baby is due in mid-February (giving an interval of some eight months) one assumes that the child was conceived just before the wedding or on the wedding-night itself.

The first child is a daughter, Katherine, and the birth is harrowingly difficult. Unusually for the period, Henry chose to be present at the delivery, 'as white (they tell me) as a sheet and the sweat dropping from his brow; and the handkerchief he had in his hand was crushed into a little ball no bigger than a musket bullet' (p. 32). Afterwards, we learn, he has difficulty in showing any tenderness towards his little daughter Katherine. Nor is the relationship between man and wife harmonious. As Ephraim tells us:

> Mrs Henry had a manner of condescension with him [Henry], such as (in a wife) would have pricked my vanity into an ulcer; he took it like a favour. She held him at the staff's end; forgot and then remembered and unbent to him, as we do to children; burthened him with cold kindness; reproved him with a change of colour and a bitten lip, like one shamed by his disgrace; ordered him with a look of the eye, when she was off her guard; when she was on the watch, pleaded with him for the most natural attentions, as though they were unheard-of favours. (pp. 31–2)

On 7 April 1749 what little serenity the Durie household enjoys is shattered by the arrival of Francis Burke, with the appalling news that James is not, after all, dead but alive in France. 'The seductive Miss Alison', as Burke provocatively calls her, faints when she hears the news (p. 34). There follows the long leeching of money from the estate to the wastrel across the water. News of James's survival also brings about a palpable change in Mrs Henry, what Ephraim calls 'a certain deprecation towards her husband' (p. 74). He, on his part, is tormented by what he conceives to be her 'truant fancies'—that is, a continuing love for James. What is most significant is that there are no more children after Katherine. Ephraim several times alludes to 'estrangement' and, the reader presumes, sexual relations between husband and wife have been suspended. So it goes on for seven years, during which period the estate is bled of £8,000.

On 7 November 1756 James returns. His family still fondly imagine that he is a political exile, in danger of his life should he be discovered. 'Mister Bally', as he is called, taunts and insults Henry from their first encounter. He reminds Ephraim of a cat playing with its prey. Alison, who is visibly affected by the reappearance of her old lover, at first tries to avoid him. Initially, he too seems inclined to keep her at a distance. But, after a furious quarrel with Henry about his refusal to dismiss Ephraim (for refusing, on his part, to drive away the pertinacious trollop Jessie Broun), James's attitude to Alison changes. She, it appears, becomes instrumental in his grand scheme of revenge against his brother. Up to that hour, Ephraim records:

> the Master had played a very close game with Mrs Henry; avoiding pointedly to be alone with her, which I took at the time for an effect of decency, but now think to be a most insidious act ... Now all that was to be changed; but whether really in revenge, or because he was wearying of Durrisdeer, and looked about for some diversion, who but the devil shall decide. (pp. 94–5)

There now begins what Ephraim calls 'the siege of Mrs Henry'. James woos her with romantic tales, ballads, and kindness towards little Katherine (whom her father seems unfairly to neglect). 'Presently there came walks in the long shrubbery, talks in the Belvedere, and I know not what tender familiarity' (p. 96). Ephraim may choose not to know, but the reader can easily guess. Alison's infatuation with James is only momentarily cooled by the revelation that he is not—after all—a Jacobite refugee, but a government spy who can move without let or hindrance between France and England. All the 'Mr Bally' business was a sham.

Ephraim is convinced that Alison is 'playing very near the fire' (p. 103) in her intimacy with James. Dark suspicions have formed in his mind, and in those of Mr Henry—so dark, in fact, that neither man can dare to articulate them:

> There were times, too, when we talked, and a strange manner of talk it was; there was never a person named, nor an individual circumstance referred to; yet we had the same matter in our minds, and we were each aware of it. It is a strange art that can thus he practised: to talk for hours of a thing, and never name nor yet so much as hint at it. And I remember I wondered if it was by some such natural skill that the Master made love to Mrs Henry all day long (as he manifestly did), yet never startled her into reserve. (p. 103)

It is, at this point, 26 February 1757, as we learn from Ephraim's pedantic notation of such matters. The events of the following night—'the fatal 27th'—occupy a whole chapter. The evening's tragedy begins with a game of cards, and too much liquor taken by James, who delivers himself of a 'stream of insult' against his luckless brother, 'Jacob' (as he tauntingly persists in calling him). There ensues the crowning insult:

> 'For instance, with all those solid qualities which I delight to recognise in you, I never knew a woman who did not prefer me—nor, I think', he continued, with the most silken deliberation, 'I think—who did not continue to prefer me'. (p. 107)

Henry strikes James, and the inevitable duel takes place in the long shrubbery. Before they cross swords, James again alludes to 'your wife—who is in love with me, as you very well know' (p. 109). In the subsequent fight James is 'killed', and the woes of the House of Durrisdeer and Ballantrae begin in earnest.

Not until much later is a crucial date introduced, almost in passing, by Ephraim:

> And now there came upon the scene a new character, and one that played his part, too, in the story; I mean the present lord, Alexander, whose birth (17th July, 1757) filled the cup of my poor master's happiness. (p. 139)

Let us review dates at this point. The Duries' first child, Katherine, was born as soon as possible after the marriage on 1 June 1748. Thereafter, for almost nine years, there were no children and—we may suspect—no sexual relations between the 'estranged' husband and wife. On 7 November 1756 James returns to the scene. He is a former lover of Alison (and may have been a lover in the carnal sense). With at least one bastard child, and the mother cruelly cast off, he is no respecter of female persons. He enjoys secret intimacies with his brother's wife in the long shrubbery and elsewhere. On 17 July, some nine months after his arrival and his 'siege' of Alison, Alexander is born. Henry, meanwhile, is so mad with sexual jealousy that he is willing to commit one of the most heinous of primal sins—fratricide.

The dates all point in one direction. So does Ephraim's coyness. It is singular that—during all the period leading up to the 'fateful 27th'—no mention is made of Alison's pregnancy. By March she will be five months pregnant. Her condition must be known—at least to Henry and Ephraim. Why do they not mention it? Because it is one of those things that they dare not even 'hint at'.

The Oxford World's Classics *The Master of Ballantrae* is edited by Emma Letley.

Oscar Wilde · *The Picture of Dorian Gray*

Why does this novel disturb us?

Early critical reaction to *The Picture of Dorian Gray* 'was almost unanimously hysterical', Isobel Murray's introduction to the World's Classics edition tells us.[1] Why? The story would seem to be excessively moralistic—a parable, no less (what shall it profit a man if he keep his good looks and lose his soul?). The homosexual hints are deeply buried beneath Dorian's conventional heterosexual villainies (the seduction of Sibyl Vane, the debauching of society wives, the ruining of young girls, the inhaling of opium). There were enough sops to the straight, late-Victorian world to have kept even Mr Pecksniff happy, one would have thought.

A key to the disturbing quality of *The Picture of Dorian Gray* can be found in the first paragraph of the novel:

> The studio was filled with the rich odour of roses, and when the light summer wind stirred amidst the trees of the garden there came through the open door the heavy scent of the lilac, or the more delicate perfume of the pink-flowering thorn. (p. 1)

This is the setting to Dorian's fateful meeting with his evil angel, Lord Henry Wotton, at the studio of his friend, the artist Basil Hallward, for whom he is sitting. The first point to note is the initial emphasis on scent. The chapter continues in the same olfactory vein, with references to Lord Henry's 'innumerable cigarettes', and the 'honey-sweet and honey-coloured blossoms of a laburnum'. These opening sentences fairly reek. The overload of smell references is compounded when we discover that the garden in which the scene is set is actually in central London, which 'roars dimly' in the background, generating its own characteristic repertoire of summer pongs.

Smells are something that the Anglo-Saxon novel is notably deficient in and uneasy about. The whole of Hemingway's fiction does not contain a single smell, as Norman Mailer once reckoned (Mailer likes a good stench in his own narratives). For most novelists in English—particularly in the nineteenth century—smells are indelicate. England has left it to its 'dirty' continental neighbours, particularly the French, to cultivate the arts of the nose in its perfume and wine industry (in which the French bibber is as attentive to aroma as taste). The two greatest novels devoted to the power of smell—Joris-Karl Huysmans' *A Rebours* and Patrick Süsskind's *Perfume*—could never have been written in England, and in translation have never been popular among English readers. Isobel Murray confirms that contemporary English critics condemned *The Picture of Dorian Gray* nose-first. 'Typical of the general outrage', she tells us, 'was an unsigned review in the *Daily Chronicle*, which condemned the novel on all counts, and chiefly as "a poisonous book, the atmosphere of which is heavy with the mephitic odours of moral and spiritual putrefaction"' (p. vii).

The Anglo-Saxon terror of smell is one likely reason why Wilde's novel triggered visceral alarm. Another is the 'unnaturalness' of the horticultural references in the first paragraph and elsewhere. In southern Britain, lilac blossoms and is odoriferous in rainy April, thorn in May ('May' is the common countryside name for it), and roses bloom in blazing mid-June (when, as we are repeatedly told by Wilde, this first chapter is set). It is not inconceivable that the flowers, blooms, and blossoms which Wilde describes (lilac, rose, laburnum, thorn) might just coincide on the branch in mid-June—but not in the full odoriferousness about which the first chapter is so eloquent and which so gratifies Lord Henry.

It is wrong—queer, one might say—that the smell of blown roses and lilac coincide and that their coincidence should be repeatedly mentioned throughout the opening scene of the novel. Sequence has been replaced by a puzzling sychronicity. It is subliminally worrying. Roses, we instinctively feel, should follow thorn and thorn should follow lilacs, they should not all merge in this sensory riot, this nasal orgasm. So too, age should follow youth. But one of the endeavours of homosexual love, with its cult of the marvellous boy, is to abolish sequence. This desire to abolish the generational sequences of youth, maturity, and age is allegorized in *The Picture of Dorian Gray*—the hero is, at one and the same time, a beautiful boy (the same object that Hallward worships in the first scene), a mature man of the world (cultivated, a member of clubs, a brilliant conversationalist at dinner tables), and a withered old man (a senility recorded in the portrait secreted in his attic). *The Picture of Dorian Gray* fantasizes a world where middle-aged hedonists can be forever boys, equated in a timeless plane composed half out of lust, half out of the wish-fulfilling visions of the fairy story. Dorian Gray is, to play with the word, two kinds of fairy—the Faustian hero who sells his soul for youth, and the middle-aged, mutton-dressed-as-lamb gay, who *would* sell his soul to look young again.

This denial of sequence also operates in the larger frame of the story's chronology. It is clear that the first scene and the narrative up to the death of Sibyl take place in the 1890s. There are unmistakeable cultural references locating the action in this period. For instance: the cult of Wagner and Schumann; the young dandies' taste for 'vermouth and orange bitters' (p. 73); the 'Yellow Book' that Dorian meaningfully glances at (p. 125); the fashion for marrying American heiresses; the 'seventeen photographs' of Dorian in different poses which Lord Henry has acquired; the repeated references to the writings of Walter Pater with which the young men first became acquainted at Oxford. The annotations to the World's Classic edition clearly indicate the time of the narrative's early action as coinciding with the period of Wilde's writing the novel, 1888–90. Looking back through literary history, we recognize it as the brief and soon-to-be curtailed golden age of aestheticism, a florescence that Wilde's own disgrace would extinguish five years later.

There are, additionally, precise chronological markers, putting the action in the late 1880s-to-1890 period. In Chapter 9 Hallward desperately tries to persuade Dorian to alter his increasingly dissipated ways. And he asks for the portrait back as he means to exhibit it, giving as his reason, 'Georges Petit is going to collect all my best pictures for a special exhibition in the Rue de Sèze, which will open the first week in October' (p. 112). Petit's gallery (which is clearly well established at this point in the narrative) was set up in 1882, in the Rue de Sèze. This must, logically, be happening at some later point in the 1880s. A little later in the novel, in Chapter 10, Dorian comes across a work of literature which will change his life:

> It was a novel without a plot, and with only one character, being, indeed, simply a psychological study of a certain young Parisian, who spent his life trying to realize in the nineteenth century all the passions and modes of thought that belonged to every century except his own, and to sum up, as it were, in himself the various moods through which the world-spirit had ever passed, loving for their mere artificiality those renunciations that men have unwisely called virtue, as much as those natural rebellions that wise men still call sin. The style in which it was written was that curiously jewelled style, vivid and obscure at once, full of *argot*

> and of archaisms, of technical expressions and of elaborate paraphrases, that characterizes the work of some of the finest artists of the French school of *Symbolistes*. (p. 125)

This novel is, unmistakably, Huysmans' *A Rebours*, which was first published in 1884.

Not to labour the point, the early chapters of *The Picture of Dorian Gray* are clearly signalled to take place in 1889–90 or, at most, a couple of years earlier. To place the action much earlier than that would be a gross misreading of the novel. How, then, do we make sense of the encounter 'eighteen years later' (as we are told) between Dorian and Sibyl's vindictive sailor-brother, Tom? Dorian escapes the knife meant for his breast only by showing his still-young face. He cannot be the 40-year-old man Tom is looking for, possessing as he does the unblemished features of someone half that age. Dorian's quick-thinking ruse is plausible, and makes an effective *coup de théâtre*. But, if we calculate, the last chapters of the novel must—given the passage of eighteen years—take place in 1903–5 at the earliest. After, that is to say, the death of Queen Victoria, the disgrace and death of Oscar Wilde, and the anathematization of the aestheticism of which Dorian is the flower. It is also a decade-and-a-half after the publication of *The Picture of Dorian Gray* in 1891.

Clearly the last chapters of Wilde's novel are not visionary glimpses of the future. The season has changed (it is murky midwinter, opposed to the high midsummer of the first chapter) but the date has not. It is still, we apprehend, 1889–90. Sequence has again been abolished. It is—as with the out-of-order flowers in the first paragraph—very disturbing. And this disturbance, I suggest, working seismically beneath the surface of what is, in most respects, a hyper-moralistic narrative, provokes those shudders which early critics and readers felt, and their invincible suspicion that there was something very queer indeed about *The Picture of Dorian Gray*. Reviewers were right to suspect that the novel's author was going to come to a very sticky end.

The Oxford World's Classics *Dorian Gray* is edited by Isobel Murray.

Thomas Hardy · *Tess of the d'Urbervilles*

Is Alec a rapist?

There has been an interesting slippage in critical discussion of the climax of 'Phase the First' ('The Maiden') of Hardy's *Tess of the d'Urbervilles: A Pure Woman* over the last century. Victorian critics, to a man and woman, assume that the luckless maiden is at least partly the author of her own misfortune. As Mowbray Morris put it in the *Quarterly Review* (April 1892):

For the first half of his story the reader may indeed conceive it to have been Mr Hardy's design to show how a woman essentially honest and pure at heart will, through the adverse shocks of fate, eventually rise to higher things. But if this were his original purpose he must have forgotten it before his tale was told, or perhaps the 'true sequence of things' was too strong for him. For what are the higher things to which this poor creature eventually rises? She rises through seduction to adultery, murder and the gallows.[1]

Writing a month earlier in *Blackwood's Magazine*, Mrs Oliphant is fierce against Tess for not having withstood temptation better:

We have not a word to say against the force and passion of this story. It is far finer in our opinion than anything Mr Hardy has ever done before. The character of Tess up to her last downfall . . . is consistent enough, and we do not object to the defiant blazon of a Pure Woman, notwithstanding the early stain. But a Pure Woman is not betrayed into fine living and fine clothes as the mistress of her seducer by any stress of poverty or misery; and Tess was a skilled labourer, for whom it is very rare that nothing can be found to do. Here the elaborate and indignant plea for Vice, that is really Virtue, breaks down altogether.[2]

Morris and Oliphant take it for granted that Tess was 'seduced'—that is, led astray, not violated or forced into sexual intercourse against her will. Compare this Victorian view with the downrightness of Tony Tanner, writing in 1968:

> Hardy's vision is tragic and penetrates far deeper than specific social anomalies. One is more inclined to think of Sophocles than, say, Zola, when reading Hardy . . . Tess is the living demonstration of these tragic ironies. That is why she who is raped lives to be hanged.[3]

She who was seduced in 1892 is she who is raped in the permissive 1960s. Even those modern commentators unwilling to go the whole way hedge their bets. Thus, the first edition of the *Oxford Companion to English Literature* (1932) declares 'Tess is seduced'. Margaret Drabble's fifth edition of *OCEL* (1984), while retaining the substance of the *Tess* entry, states 'Tess is cunningly seduced'. The 1993 literary encyclopaedia *The 1890s* backs both horses by opening its *Tess* entry with the statement: 'This simple story of seduction-cum-rape', and goes on to describe what happens in the Chase as 'virtual rape'.[4] Ian Gregor, in his influential study of Hardy's major novels, goes all the way by declaring that 'it is both a seduction *and* a rape' (try that in court).[5] The World's Classics editors, Juliet Grindle and Simon Gatrell, use the ambiguous term 'betrayed' (as does OUP's blurb-writer) and the non-felonious 'violated' ('violation' will not land you with a ten-year prison term, 'rape' will). Writing in the 1990s, with the complexities of 'date rape' hovering in the air, James Gibson refers edgily to Tess's 'sexual molestation by Alec' and his 'sexual harassment of his victim'. In his outline of the crucial episode in the 'Text Summary' section of the

'Everyman' edition, Gibson reverts to a more Tannerian reading of events:

Chapter 11
Alec rides with Tess into the Chase—works on her by appearing to be worried about her safety and emphasising the presents he has sent to her family—he deliberately loses the way—after pretending to go in search of the way he returns and rapes her.[6]

So what does happen in the Chase on that September night—seduction, cunning seduction, betrayal, sexual molestation, sexual harassment, violation, 'virtual rape', or rape? The first point to note is that Hardy himself was somewhat unsure about the 'naughty chapters' (as Mrs Oliphant called them). In the serialization of the story in the *Graphic* newspaper (July–December 1891) he was prevailed on by nervous editors to drop the Chaseborough dance and subsequent seduction/rape sequence altogether, putting in its place an entirely new sub-plot. In the *Graphic* version of *Tess*, the heroine is tricked into a fake marriage, and is thus deflowered with her full (if deluded) consent. This venerable device had been used earlier by Charles Reade in *Put Yourself in his Place* (1870), by Thackeray in *Philip* (1862), and, aboriginally, by Scott in *St Ronan's Well* (1823). The attraction of the bogus-marriage gimmick was that it enabled the hoodwinked heroine to commit the act of fornication innocently, thus preserving her 'purity'. In the three-volume edition, which came out in November 1891, Hardy repudiated artifice and insisted on reprinting the original rape/seduction text, with the prefatory proclamation: 'If an offence come out of the truth, better is it that the offence come than that the truth be concealed.'

Even in the frank versions of 'The Maiden', however, much is left inscrutable, if not entirely concealed. The rape/seduction episode begins with a description of the Trantridge peasantry's loose morals and hard drinking, which are given free rein in their Saturday night festivities at the nearby 'decayed market-town', Chaseborough. Tess, we are told, likes to go to these Saturday night affairs, although she does not participate in the revelry. On the misty September Saturday night in question, Tess makes her way from Chaseborough to a barn in a nearby 'townlet' where her fellow Trantridge cottagers are at a 'private jig'. She wants their company on the way home, since there has been both a fair and a market day at Chaseborough, and there may be drunken men in the country lanes. When she arrives at the dance, Tess discovers a surreal scene. The barn floor is deep in 'scroff'—'that is to say the powdery residuum from the storage of peat and other products, the stirring of which by [the dancers'] turbulent feet created [a] nebulosity that involved the scene . . . [a] floating, fusty *débris* of peat and hay' (p. 66). The dusty haze thrown up by the dancers' muffled stamping merges with the mist, and later the fog, which enshrouds the whole of the seduction/rape episode in a corresponding moral 'nebulosity'.

The dustily indistinct picture of the dance has been connected with the pictorial influence of French Impressionism on Hardy and is one of his fine visual set-pieces. At a more physical level, the scene alludes to a common belief in country communities—that flying dust, as it gets trapped in their underwear, has a sexually exciting effect on women dancers. It is part of the folklore of barn-dances in America that unscrupulous young men—intending to induce wantonness in their partners—scatter pepper on the boards before the evening gets under-way. Certainly the 'scroff' seems to have had an aphrodisiac effect at Chaseborough. Hardy hints at sexual orgy by a string of meaningful classical references:

They coughed as they danced, and laughed as they coughed. Of the rushing couples there could barely be discerned more than the high lights—the indistinctness shaping them to satyrs clasping nymphs—a multiplicity of Pans whirling a multiplicity of Syrinxes; Lotis attempting to elude Priapus, and always failing. (pp. 66–7)

Each of these three allusions signals 'rape'. Tess is invited to join in by a dusty, sweating swain. But she refuses. She becomes aware of the glowing tip of Alec's cigar in the gloom behind her. He offers to take her home, but although she is very tired she declines, not quite trusting him—perhaps forewarned by the phallic heat of his Havana on her neck. But later, when Car Darch (one of Alec's cast mistresses) threatens violence, Tess allows herself (a maiden in distress) to be rescued by Alec, now on horseback. This is the prelude to Chapter 11. Chaseborough is only three miles from Trantridge, and the journey on Alec's stallion should take twenty minutes or so. But he deliberately loses his way turning his horse into the foggy wilderness of the Chase—'the oldest wood in England'. Tess, who has been up since five every day that week, is exhausted. As she falls asleep in the saddle, Alec puts his arm around her waist. 'This immediately put her on the defensive,' we are told, 'and with one of those sudden impulses of reprisal to which she was liable she gave him a little push from her' (p. 74). Alec almost tumbles off the horse. It is a significant detail. The push looks forward to Tess's eventually stabbing Alec to death, which—we apprehend—is another reflexive 'impulse of reprisal'. Here it stresses that even when her body is dormant, Tess's purity is vigilant and well capable of defending itself. This is important, since she will be sleeping when the seduction/rape occurs.

The point is also stressed that Alec has not directed his horse's head into the Chase with any overtly mischievous intention, but merely 'to prolong companionship with her' (p. 76). Tess repulses his love-making as they ride, without ever distinctly denying that she loves him. He is much encouraged by her lack of 'frigidity'. He contrives further to weaken her resolve with the information that he has bought her father a new horse, to replace the luckless Prince, who died as a result of her falling asleep on the road. Once again, it seems that Tess is failing into dangerous slumber. Alec, who by now is completely lost in the fog and trees, wraps her in his coat, makes a 'sort of couch or nest' for her in the newly dropped leaves, and goes off on foot to look for some landmark. He eventually locates the road, and returns to find Tess fast asleep. He bends down to her, 'till her breath warmed his face, and in a moment his cheek was in contact with hers. She was sleeping soundly, and upon her eyelashes there lingered tears' (p. 77).

This is the last image Hardy leaves us with. It could be Prince Charming about to wake Sleeping Beauty, or it could be ravishing Tarquin. The narrative averts its gaze from whatever happens next and moralizes loftily for three paragraphs. The clearest clue as to what is meanwhile going on between Tess and Alec is given in the second of these paragraphs:

> Why it was that upon this beautiful feminine tissue, sensitive as gossamer, and practically blank as snow as yet, there should have been traced such a coarse pattern as it was doomed to receive; why so often the coarse appropriates the finer thus, the wrong man the woman, the wrong woman the man, many thousand years of analytical philosophy have failed to explain to our sense of order. One may, indeed, admit the possibility of a retribution lurking in the present catastrophe. Doubtless some of Tess d'Urberville's mailed ancestors rollicking home from a fray had dealt the same measure even more ruthlessly towards peasant girls of their time. But though to visit the sins of the

fathers upon the children may be a morality good enough for divinities, it is scorned by average human nature; and it therefore does not mend the matter. (p. 77)

As Thackeray says in *Vanity Fair*, the novelist knows everything. Hardy must know what is going on here, even if he chooses not to tell us. Clearly intercourse is taking place while the narrator turns away and prates about olden times. But what kind of intercourse? All the narrative divulges to the reader is that Alec is not as 'ruthless' as those ancient ravishers, Tess's ancestors, taking their seigneurial rights. It is clear that Alec has not set out with the explicit purpose of assaulting Tess; when he leaves her in her leafy couch, it is genuinely to find the road home, not to lull her into defencelessness. Not that she would normally be defenceless. The point is made earlier in the chapter that even when asleep, Tess is able to fend off unwanted sexual advances. Why does she not protect her imperilled virtue with one of those timely 'impulses of reprisal'?

More significantly, when—a maiden no more—Tess upbraids Alec, she does not accuse him of rape, but of having duped her: 'I didn't understand your meaning till it was too late' (p. 83), she says. Nor, when upbraiding her mother for not warning her against men, does Tess claim that she has been raped. As the narrative glosses her thoughts:

She had never wholly cared for [Alec], she did not at all care for him now. She had dreaded him, winced before him, succumbed to adroit advantages he took of her helplessness; then, temporarily blinded by his ardent manners, had been stirred to confused surrender awhile: had suddenly despised and disliked him, and had run away. (p. 87)

Alec, we understand, has been 'adroit'—some cunning caresses with his hands are implied. His 'ardent manners' (an odd conjunction—ardour is rarely well mannered) had 'stirred' Tess—erection is hinted at. The verb 'stirred' is significant, suggesting as it does physical reciprocation on Tess's part. Did she consent? 'Confused surrender' suggests that she did, but that she was blinded at the time by his stimulating foreplay and the power of her own aroused feelings.

By Victorian legal lights it was clearly seduction; there was nothing forcible in Alec's actions, although, as he himself avows ('I did wrong—I admit it'), they were the actions of a cad. He 'took advantage' of her. This is immoral, but not criminal. Even by the strict 1990s definition of rape on North American campuses, his behaviour would probably not be criminal. It is not recorded that Tess clearly told Alec to stop, once he had started to make love to her.' Stirred' as she was, she may well have encouraged him to continue making love by body movements of her own. That is, neither seduction nor rape may be the proper term; Tess was a willing, if misguided, participant in her own undoing.

Why then do modern critics and readers assume that Alec is a rapist? For the same reason that they are unwilling to see Tess as a murderess. Here again, Hardy manipulates our response in his heroine's favour less by what he describes than by what he omits to describe. To summarize: hearing unusual 'sounds', Mrs Brooks, the landlady at the Herons, the inn where Alec and Tess have taken an apartment, looks through the keyhole. She sees Tess in distress at the breakfast table, and hears a long complaint ('a dirge rather than a soliloquy') from her lips. Tess is berating herself for her weakness in surrendering again to Alec's 'cruel persuasions'. She sees herself as an irredeemably fallen adulteress. Mrs Brooks hears 'more and sharper words from the man', then 'a sudden rustle'. Tess soon after hurries away from the inn, dressed in black. Alec's body is discovered on the bed, stabbed through the heart (pp. 367–70).

Tess's subsequent explanation to Angel is not entirely satisfactory.

> 'But how do you mean—you have killed him?'
>
> 'I mean that I have,' she murmured in a reverie.
>
> 'What—bodily? Is he dead?'
>
> 'Yes. He heard me crying about you, and he bitterly taunted me; and called you by a foul name; and then I did it; my heart could not bear it: he had nagged me about you before—and then I dressed myself, and came away to find you.' (p. 372).

Hardy does not give us any details of the subsequent trial, leaping straight from Tess, arrested on the sacrificial slab at Stonehenge, to her execution at Wintoncester Gaol. But it would be interesting to know what came out in court. By Mrs Brooks's testimony (as we have it), it would seem that Alec—justifiably vexed by Tess's long diatribe against himself—said something 'sharp' ('bitterly taunted' seems an overstatement). It is hard to think up a 'foul name' applicable to a man ('no-balls eunuch'?) which would justify what followed. Tess then picked up the carving knife from the breakfast table, walked all the way across the length of the living room to the bed on which Alec was still lying, and stabbed him through the heart, One precisely aimed stroke has killed him. It is hard to imagine how this could be done—given Alec's superior strength and agility—unless Tess waited until he relapsed into sleep, as people do before breakfast. An awake Alec would hardly watch Tess stalking towards him with an upraised knife without raising a hand to defend himself or shifting his torso away from the path of the murder weapon.

It is conceivable that a legal defence could be made for Tess along the lines that lawyers successfully defended Lorena Bobbit—the aggrieved Virginian who cut off her husband's penis with a carving knife while he slept.

Possibly a 1990s jury might acquit Tess, on the grounds that she, like Lorena, had suffered years of abuse from her partner (although Alec is not recorded as ever striking or beating Tess). But the jury would not, one imagine, acquit her as readily and absolutely as do the literary critics, This, for instance, is how James Gibson summarizes the climactic chapter:

Chapter 56
The landlady of the lodging-house is curious—hears Tess moaning in her room and sharp words—she sees Tess leaving the house and then a red spot on the ceiling—Alec is dead.[7]

'Alec is murdered' would seem to be truer to the facts. Hardy's rhetoric allows the critic to overlook the simple wrongness of Tess's act, and mask it in a neutral phraseology more appropriate to suicide or death by natural causes than homicide. The holes in Hardy's account allow us to jump to conclusions ('Alec is a rapist who gets what is coming to him') and sanction such exonerating imagery as Tony Tanner's: 'Tess is gradually crucified on the oppugnant ironies of circumstance and existence itself'. On objective legal grounds, one might retort, Tess deserves crucifixion rather more than do the two thieves and their famous companion. It would be much harder to sustain the 'Tess as Christ figure' line if readers had before them a clear image of her plunging the carving knife into Alec's sleeping body, choosing her spot carefully so as to kill him with one blow—all because he had applied some unspecified 'foul name' to her husband. Nor would it be easy, I imagine, to sustain the 'Tess is a victim of rape' line (which leads directly into the 'justifiable homicide' line) if one had a clear image of her making reciprocal love to Alec in the Chase. Hardy's novel is like a court case in which all the material evidence is left out, and the jury (readers) rushed to judgement on the basis of the defendant's beauty and pathetic suffering alone.

All this is not to suggest that Tess is a murderous slut who gets what is coming to her. But one should perhaps give more credence to Mrs Oliphant's Victorian commonsensical view. Alec is not a rapist and, although her innocence makes her vulnerable, Tess must take some small responsibility for what happens in the Chase. Tess does have a saleable skill, and she did not have to surrender a second time to Alec purely for economic reasons. Nor, having surrendered, did she have to compound adultery with wilful murder. Had Hardy, in the manner of some Victorian John Grisham, supplied us with two closely described trials in the body of the novel (the first of Alec for rape, the second of Tess for murder), our verdict would surely be harder on Tess, and lighter on Alec.

The Oxford World's Classics *Tess of the d'Urbervilles* is edited by Juliet Grindle and Simon Gatrell, with notes by Nancy Barrineau.

Arthur Conan Doyle · *Sherlock Holmes*

Mysteries of the Speckled Band

'The Adventure of the Speckled Band' is one of Sherlock Holmes's best-known early cases.[1] The affair is recalled by Watson, much later in life. The lady who initially implored the great detective's assistance has, we guess, recently died (prematurely, it would seem), allowing Watson to divulge her story to the world. As Watson recalls, in April 1883 Miss Helen Stoner, a handsome but clearly distressed young woman with prematurely white strands in her hair, called at 221B Baker Street, early in the morning. Her tale was intriguing. Raised in India, she had returned to England some ten years earlier with her mother, her twin sister Julia, and her stepfather. The girls' mother has died eight years since, 'in a railway accident near Crewe' (p. 175). Helen's stepfather; who evidently married his wife for her money (some £1,100 a year, shrunk by current agricultural distress to £750), is an unredeemable brute, the degenerate offspring of a 'dissolute and wasteful' family on whose decayed estate, Stoke Moran in Surrey, they now live in reduced circumstances. A doctor by profession, Roylott is suspected of having murdered a servant in India and—as Holmes astutely deduces—routinely brutalizes his stepdaughters (the detective's sharp eyes note bruises on Miss Stoner's wrist and arm). The Misses Stoner, on marriage, will take to their husbands a portion of their mother's inheritance. Julia, however, has died very recently in suspicious circumstances after indicating an intention to marry.

Helen describes the enigmatic events surrounding her

sister's death in some detail. On the evening she died, the young ladies, as had happened before, heard a 'low whistle', for which they had no explanation. There was also heard a clanging sound, which was similarly mysterious to them. The girls have separate bedrooms and, after retiring, Helen was roused by a hideous shriek. She rushed out to encounter Julia emerging from her room, 'swaying like a drunkard'. The unfortunate woman convulsed and died seconds later, in agony. There were no marks apparent on her body, nor any evident cause of death. Julia's final words are baffling in the extreme:

> as I bent over her she suddenly shrieked out in a voice which I shall never forget, 'Oh, my God! Helen! It was the band! The speckled band!' There was something else which she would fain have said, and she stabbed with her finger into the air in the direction of the Doctor's [i.e. her stepfather's room], but a fresh convulsion seized her and choked her words. (p. 178)

The doctor, it emerges, has lately been intimate with a band of gipsies on his 'plantation'. These are the only human beings with whom he has any civilized intercourse. Possibly, conjectures Miss Stoner, the mysterious speckled band may refer 'to the spotted handkerchiefs which so many of them [i.e. gipsies] wear over their heads' (p. 180). But since no one has been seen entering or leaving the room, what we would seem to have here is a version of the 'locked room mystery'—that master-class problem for all great sleuths.

These, Holmes opines, 'are very deep waters'. Helen goes on to say that she too is about to be married, to 'Percy Armitage—the second son of Mr Armitage, of Crane Water, near Reading' (p. 180). Her stepfather has offered no objection to the match, but she has anxieties. More so, since she has been moved by his instruction into the very bedroom that her unlucky sister formerly occupied, and has the previous night heard the same ominous whistle that they heard on the night of Julia's death. After Miss Stoner has left, Holmes and Watson receive an unpleasant call from Dr Grimesby Roylott, of Stoke Moran. He confirms his stepdaughter's unflattering account. An old man possessed of almost superhuman strength (he bends a poker double to add emphasis to his threats), his face is 'marked with every evil passion'. He utters some inarticulate but very horrible imprecations against Holmes, should he interfere.

Holmes later admits that at this initial stage he had formed 'an entirely erroneous conclusion, which shows, my dear Watson, how dangerous it always is to reason from insufficient data'. The presence of the gypsies, and 'the use of the word "band"' had thrown the sleuth on an 'entirely wrong scent' (p. 196). We never discover what Holmes's erroneous 'conclusion' was—presumably that Roylott summoned gipsy assassins to his daughter's chamber for light-footed murder.

After the visits of Miss Stoner and Dr Roylott, Holmes and Watson take the train down to Stoke Moran. They discover in the dead girl's bedroom a dummy bell-pull. The bed is bolted to the floor, so that it must always be directly underneath the tasselled cloth band, which is odd. Holmes's sharp eyes also detect an air-vent which has been recently constructed. Its passage does not connect, as would be logical, to an outside wall, but to the doctor's adjoining bedroom. This vent allows the rank odour of Roylott's cigar and even gleams of light to penetrate into Helen's chamber. It is not said outright, but we deduce that the vent could also serve as a spy-hole for anyone with voyeuristic intentions. Dr Roylott, of course, is no gentleman and a young lady's privacy would not be sacred to him.

In every way Stoke Moran gives Holmes and Watson the impression of being an 'evil household'. The outlying estate is similarly infested with signs of evil. 'There was little difficulty in entering the grounds,' Watson recalls:

> for unrepaired breaches gaped in the old park wall. Making our way among the trees, we reached the lawn, crossed it, and were about to enter through the window, when out from a clump of laurel bushes there darted what seemed to be a hideous and distorted child, who threw itself upon the grass with writhing limbs, and then ran swiftly across the lawn into the darkness.
>
> 'My God!' I whispered; 'did you see it?'
>
> Holmes was for the moment as startled as I. His hand closed like a vice upon my wrist in his agitation. Then he broke into a low laugh, and put his lips to my ear.
>
> 'It is a nice household,' he murmured. 'That is the baboon.' (p. 193)

Baboons, of course, are not native to Surrey. This beast is a relic from an menagerie of Indian animals which the doctor tried unsuccessfully to set up on his return from the subcontinent. But, for astute readers of detective stories, the introduction of the baboon will recall that archetype of the locked-room mystery genre, Edgar Allan Poe's 'The Murders in the Rue Morgue' (1841), where the murderer is finally revealed to be an orang-utan. Might the baboon have somehow been responsible for Julia's death?

The denouement is quickly told. In a metal safe (with a sonorously clanging door) in his bedroom Doctor Roylott keeps a trained swamp adder—the 'deadliest snake in India'. This beast has been trained to respond to whistled signals. It is capable of slithering through the connecting vent and down the fabric bell-rope. (It defies plausibility, incidentally, that it could slither back up—some variant of the Indian rope trick was apparently in Doyle's mind.)[2] This serpent is the 'speckled band' of which the inarticulate and dying Julia spoke. Surprised by Holmes and Watson as it slithers towards its second victim, the infuriated snake insinuates its way up the bell-pull and back into Roylott's bedroom and stings him to death. Holmes has, quite advertently as he later confesses, engineered the doctor's death. 'I cannot say that it is likely to weigh very heavily on my conscience' (p. 197), he complacently informs Watson.

There remain, even after the thrilling denouement, profound mysteries to this case. The deceased Miss Stoner was, like her twin sister Helen, brought up in India where she spent what we can calculate to be her first fifteen years or so of life. This experience would certainly have educated her as to the existence of snakes. Why, then, after having been agonizingly stung and clearly having seen what stung her, does she waste her last words talking about 'speckled bands' as if she had never seen such a thing as a swamp adder, or its serpentine cousins? Even if an old India hand might have had a momentary confusion on waking from sleep, it would quickly have been dispelled by the sinuous movements of the 'band', and its bite. That bite, incidentally, is anything but painless. The Doctor emits a 'dreadful shriek' on being bitten which is audible in the nearby village. One might also carpingly note that, although a swamp adder might be trained to return to its nest on command, it could scarcely be trained to bite on command.

When one looks more closely at it, the whole question of what it is Julia sees is vexed. According to Helen, when she ran out, alarmed by Julia's 'wild scream', into the passage connecting their bedrooms: 'By the light of the corridor lamp I saw my sister appear at the opening, her face blanched with terror, her hands groping for help, her whole figure swaying to and fro like that of a drunkard' (p. 178). As Helen puts her arms around the dying woman, 'she suddenly shrieked out in a voice which I shall never forget, "O, my God! Helen! It was the band! The speckled band!"' At this point, she falls speechless, although as

Helen notes, 'there was something else she would fain have said, and she stabbed with her finger into the air in the direction of the Doctor's room' (p. 178).

While writing the scene Conan Doyle evidently realized that, in order for the victim to have seen the 'speckled band', there must be some source of light in Julia's room. If there were too much light she would naturally see the reptile slithering up and down the bell-pull, and make too correct an identification. Thus, in reply to Holmes's question, 'Was your sister dressed?', Helen is made to reply, 'No she was in her nightdress. In her right hand was found the charred stump of a match, and in her left a matchbox' (p. 179). The feeble light afforded by a Victorian Lucifer would explain how a snake might look like a ribbon. But how, one may ask, when she emerged into the corridor did Julia's hands 'grope for help', or her finger point towards the doctor's room?[3]

Why, as a further mystery, did Julia not complain at the doctor's knocking a hole into the adjoining wall between their bedrooms, violating her privacy and adding nothing to the freshness of her boudoir? Why do the girls not leave Stoke Moran? There is no reason for Julia to have remained there: she was of age, and her prospective in-laws would certainly have given her a haven (as, of course, would the Armitages of Reading for Helen, were she to ask them). There is, one deduces, a strong suggestion of sexual bondage. The girls, for some reason, cannot leave. We never learn how Helen came by the bruises on her wrists that Holmes discerns at their first interview in Baker Street. All that she says, by way of blushing explanation, is that 'He is a hard man ... and perhaps he hardly knows his own strength' (p. 181). This, notoriously, is what battered wives tell the police and magistrates in defence of their brutal husbands. 'Miss Roylott', says Holmes, 'You are screening your stepfather' (p. 181). Why, in saying this, does Holmes, who has the best brain in England, miscall Miss Stoner 'Miss Roylott'—is he hinting at a closer relationship than that of stepfather and stepdaughter? Has the unspeakable Dr Roylott somehow mixed his identity with that of the young woman? Helen Stoner, we note, does not correct Holmes when he calls her 'Miss Roylott'—is he not subtly probing for evidence of incest, and has he not found it?[4]

This whole episode recalls a vivid exchange between Holmes and Watson in another exciting early case, 'The Adventure of the Copper Beeches'. As they travel into rural Hampshire Watson enthuses about the beauty of the scene. 'Good heavens!', he exclaims, 'Who would associate crime with these dear old homesteads?' Holmes replies bleakly that such idyllic landscape fills him with more 'horror' that 'the lowest and vilest alleys in London':

> the reason is very obvious. The pressure of public opinion can do in the town what the law cannot accomplish. There is no lane so vile that the scream of a tortured child, or the thud of a drunkard's blow, does not beget sympathy and indignation among the neighbours, and then the whole machinery of justice is ever so close that a word of complaint can set it going, and there is but a step between the crime and the dock. But look at these lonely houses, each in its own fields, filled for the most part with poor ignorant folk who know little of the law. Think of the deeds of hellish cruelty, the hidden wickedness which may go on, year in, year out, in such places, and none the wiser. (p. 280)

Reading 'The Adventure of the Speckled Band' at a deeper level than the sleuth's slick cracking of the case, there remains a surplus of unresolved questions. Why did Helen, after being released from Stoke Moran, die so prematurely? Was it some ineradicable legacy of shame? Why does Holmes feel justified in killing Dr Roylott as justifiably as one might squash a dung beetle? He is not normally given to such acts of vigilantism. What 'hidden wickedness' may we surmise was going on at Stoke Moran under the doctor's corrupt regime? Much more, we apprehend, than the amiably dull-witted Watson suspects.

The Oxford World's Classics *The Adventures of Sherlock Holmes* is edited by Richard Lancelyn Green.

Thomas Hardy · *Jude the Obscure*

What does Arabella Donn throw?

Most readers will, if they have learned nothing else from *Jude the Obscure*, know how a pig of the nineteenth century was killed, dressed, and its carcass disposed of. 'The pig-sticking and so forth', as D. H. Lawrence dismissively calls it in his *Study of Thomas Hardy*, is uncomfortably prominent. Hardy's proclaimed intention in *Jude* was to tell his story 'without a mincing of words'. Where things porcine are concerned, his words are notably unminced. There are, however, some aspects of this strand in the novel which may elude the modern reader, and are perhaps designed by Hardy to be elusive. Take, for example, the irruption of Arabella into Jude's life, with the 'missile':

> In his deep concentration on these transactions of the future [Jude is dreaming of becoming a bishop], Jude's walk had slackened, and he was now standing quite still, looking at the ground as though the future were thrown thereon by a magic lantern. On a sudden something smacked him sharply in the ear, and he became aware that a soft cold substance had been flung at him, and had fallen at his feet.
>
> A glance told him what it was—a piece of flesh, the characteristic part of a barrow-pig, which the countrymen used for greasing their boots, as it was useless for any other purpose. (p. 35)

What strikes one here is the odd tentativeness which masks an obvious knowledgeability about country matters. The parodic cupid's dart is described with the maximum of periphrasis compatible with not actually disguising what the organ is, 'a piece of flesh, the characteristic part of a barrow-pig'. Twice more in the scene it is referred to. Arabella saucily observes: 'If I had thrown any thing at all, it shouldn't have been *that*' (p. 35). Later on it occurs to the otherwise gullible Jude that 'it had been no vestal who chose *that* missile for opening her attack on him' (p. 39). In neither case is 'that' specified any more than it was on its first appearance.

Hardy's delicacy of description serves a number of purposes. First, and most simply, he could not be as frank on matters of sexual detail as he might have liked. Mrs Grundy—in the form of his editors and publishers—would not let him. More artistically, the lack of specification aptly mimics the coyness of the 'maidens' and conveys the prim disgust of Jude, who shrinks from naming the object ('*that*'), even in the private recesses of his own mind. The maidens' motive for avoiding the word is aggressive sexual mock-modesty, his is genuine sexual timidity. And, at the rhetorical level, yet another function is performed. Hardy must have been aware that relatively few of his public (townees all) would have known what part of pork offal was used for boot-dubbin in the West Country.

The reader is thus teased into supplying his or her own suggestion for what '*that*' might be. Not everyone gets it right. Kate Millett, for example, in her otherwise acute discussion of *Jude* in *Sexual Politics*, writes that the lovers 'first meet when Arabella pitches the scrotum of a butchered barrow-pig at Jude's head'.[1] This is a misapprehension which leads to a misreading of the episode. A barrow-pig is a castrated boar reared for pork, not breeding, and *sui generis*, has no scrotum. That part of the beast's sexual equipment is cut away while it is still a piglet, so that it will grow a fat, porcine eunuch. What Arabella throws at Jude is a pizzle or penis. But nor is it quite sufficient merely to describe it as a 'penis', as Patricia Ingham does, in her account of how Hardy was forced to bowdlerize this scene for its various form of publication (p. 435). The stress is on the observation 'it was useless for any other purpose'. It is a dysfunctional penis. The 'message' is not one of sexual invitation, but one of taunting. Arabella's missile is not a symbol of animal potency but of animal impotency. The message is mischievously provocative ('Is yours as useless as this, young man?') This aggressive sexual taunting is a note which will reverberate through the lovers' subsequent relationship until Arabella finally disposes of her thoroughly emasculated mate at Christminster.

The Oxford World's Classics *Jude the Obscure* is edited by Patricia Ingham.

R. L. Stevenson · *Weir of Hermiston*

What is Duncan Jopp's crime?

The Oedipal struggle between Archie Hermiston and his awesome, hanging-judge father reaches a crisis in Chapter 3—'In the Matter of the Hanging of Duncan Jopp.' The chapter opens:

It chanced in the year 1813 that Archie strayed one day into the Judiciary Court. The macer made room for the son of the presiding judge. In the dock, the centre of men's eyes, there stood a whey-coloured, misbegotten caitiff, Duncan Jopp, on trial for his life. His story, as it was raked out before him in that public scene, was one of disgrace and vice and cowardice, the very nakedness of crime; and the creature heard and it seemed at times as though he understood—as if at times he forgot the horror of the place he stood in, and remembered the shame of what had brought him there. He kept his head bowed and his hands clutched upon the rail; his hair dropped in his eyes and at times he flung it back; and now he glanced about the audience in a sudden fellness of terror, and now looked in the face of his judge and gulped. There was pinned about his throat a piece of dingy flannel; and that it was perhaps that turned the scale in Archie's mind between disgust and pity. The creature stood in a vanishing point; yet a little while, and he was still a man, and had eyes and apprehension; yet a little longer, and with a last sordid piece of pageantry, he would cease to be. And here, in the meantime, with a trait of human nature that caught at the beholder's breath, he was tending a sore throat. (p. 103)

There is a striking anticipation here of a passage in George Orwell's 'A Hanging':

It was about forty yards to the gallows. I watched the bare brown back of the prisoner marching in front of me. He walked clumsily with his bound arms, but quite steadily, with that bobbing gait of the Indian who never straightens his knees. At each step his muscles slid neatly into place, and the lock of hair on his scalp danced up and down, his feet printed themselves on the wet gravel. And once, in spite of the men who gripped him by each shoulder, he stepped slightly aside to avoid a puddle on the path.

It is curious, but till that moment I had never realized what it means to destroy a healthy, conscious man. When I saw the prisoner step aside to avoid the puddle I saw the mystery, the unspeakable wrongness, of cutting a life short when it is in full tide.[1]

As an increasingly appalled Archie watches, Jopp and his slatternly mistress, Janet, are ruthlessly mocked by the judge in his braid-Scots dialect—'Godsake! ye make a bonny couple.' Janet is spared, but Duncan is summarily sentenced to execution. In the course of passing sentence, Weir delivers himself of a particularly brutal *obiter dictum:* 'I have been the means, under God, of hanging a great number, but never just such a disjaskit [untidy] rascal as yourself'. As the narrative observes: 'The words were strong in themselves; the light and heat and detonation of their delivery, and the savage pleasure of the speaker in his task, made them tingle in the ears' (p. 104).

Duncan Jopp leaves the dock, a pathetic spectacle in the eyes of court: 'Had there been the least redeeming greatness in the crime, any obscurity, any dubiety, perhaps he might have understood'. For Archie the trial—more particularly the cosmic insignificance of Jopp as a defendant—has soiled his father irredeemably: 'It is one thing to spear a tiger, another to crush a toad; there are aesthetics even of the slaughter-house; and the loathsomeness of Duncan Jopp enveloped and infected the image of his judge' (pp. 104–5). But neither here nor elsewhere does Stevenson indicate what Jopp's 'crime' is, other than in the contemptuous vagueness of 'His story . . . was one of disgrace and vice and cowardice, the very nakedness of crime'.

The short interval between sentencing and execution passes as a 'violent dream' for Archie. He is present at the public hanging of Jopp (in Edinburgh's Grass Market, presumably):

> He saw the fleering rabble, the flinching wretch produced. He looked on for a while at a certain parody of devotion, which seemed to strip the wretch of his last claim to manhood. Then followed the brutal instant of extinction, and the paltry dangling of the remains like a broken jumping-jack. He had been prepared for something terrible, not for this tragic meanness. He stood a moment silent, and then—'I denounce this God-defying murder', he shouted; and his father, if he must have disclaimed the sentiment, might have owned the stentorian voice with which it was uttered. (pp. 105–6)

It is Archie's first overt act of rebellion against his father, and as such a threshold event in his life. He compounds his rebellion at the place of execution by proposing that evening at the 'Spec' debating society the 'Jacobinical' motion: 'Whether capital punishment be consistent with God's will or man's policy?' All this inevitably gets back to Hermiston who, with his habitual, unmanning, insolence banishes Archie to the family estate at Hermiston. The scene is thus set for the events which will lead (had Stevenson lived long enough to write it) to Archie's being brought up before his own father on a charge of murder and—presumably—being sentenced to the same fate as the luckless Jopp.

Stevenson devotes a great deal of space in Chapter 3 to the description of Duncan Jopp. But he nowhere tells us what his offence is. As is well known (and as RLS pointed out in letters to friends), Weir of Hermiston is based closely on the 'Scottish Jeffreys', Robert MacQueen, Lord Braxfield (1722–99). As Emma Letley explains, in her notes to the World's Classics edition:

Braxfield became an advocate in 1744 and, in 1766, was appointed a Lord of Session; in 1788 he became Justice-Clerk, in effect head of the Criminal Court in Scotland. He was known for being particularly harsh with political offenders, and was noted for his brutality and insulting treatment of such plaintiffs as appeared before him. (p. 215)

A 'coarse and illiterate man' (as Henry Cockburn called him) and spectacularly drunken to boot, who loved to use broad dialect on the bench, Braxfield made himself hated (and earned his 'Jeffreys' sobriquet) in the brutal repression of the Duns rioters over which he presided in 1793–4. His conduct was, Cockburn declared, 'a disgrace to the age'. But *Weir of Hermiston* is set in a later age—1813. Whatever else, Jopp is not guilty of any offence against civil order; his, we may be sure, is no 'political' crime. But what kind of crime is it? It is some two decades before the Peel reforms and Jopp could, in 1813, be hanged for the theft of a loaf of bread to feed his starving offspring. But his crime seems more sordid than this. 'Grant he was vile', Archie tells his father, 'why should you hunt him with a vileness equal to his own?' Some squalid sexual crime seems to be indicated. The point is necessary to assist the readers in arranging their sympathies towards father and son. If Jopp, for instance, had killed a man, or a woman, or a child, in cold blood we might well feel—even with the enlightened sympathies of the twentieth century—that the hanging was no 'God-defying murder', but justice. If, on the other hand, Jopp was a petty thief, or some wretched sexual delinquent, hanging would indeed strike the modern reader as judicial murder. By not instructing on this matter Stevenson, for his own artistic purposes, leaves us, like Jopp, twisting in the wind.

The Oxford World's Classics *Weir of Hermiston* is edited by Emma Letley.

H. G. Wells · *The Invisible Man*

Why is Griffin cold?

H. G. Wells was keen that his 'scientific romances' should be just that—scientific. He devotes a whole section of *The Invisible Man* (Chapter 19, 'Certain First Principles') to authentication of the central concept in the novel. It is, for the lay-reader at least, an extraordinarily plausible performance. 'Have you already forgotten your physics?', Griffin asks his old class-mate, Kemp, before launching into a lecture on optics:

> Just think of all the things that are transparent and seem not to be so. Paper, for instance, is made up of transparent fibres, and it is white and opaque only for the same reason that a powder of glass is white and opaque. Oil white paper, fill up the interstices between the particles with oil so that there is no longer refraction or reflection except at the surfaces, and it becomes as transparent as glass. And not only paper, but cotton fibre, linen fibre, woody fibre, and *bone*, Kemp, *flesh*, Kemp, *hair*, Kemp, *nails* and *nerves*, Kemp, in fact the whole fabric of a man except the red of his blood and black pigment of hair, are all made up of transparent, colourless tissue. So little suffices to make us visible one to the other. For the most part the fibres of a living creature are no more opaque than water. (p. 92)

'Of course, of course', responds Kemp, suddenly remembering his undergraduate physics, 'I was thinking only last night of the sea larvae and all jelly-fish!'

Wells managed to finesse the business about the black pigment in hair by making Griffin albino. But one feature of his invisibility hypothesis continued to bother the author, and evidently remained insolubly inauthentic.

As he later explained in a lotter to Arnold Bennett:

Any alteration of the refractive index of the eye lenses would make vision impossible. Without such alteration the eyes would be visible as glassy globules. And for vision it is also necessary that there should be visual purple behind the retina and an opaque cornea and iris. On these lines you would get a very effective short story but nothing more.[1]

There are other irrationalities flawing the central conception of *The Invisible Man* which Wells seems not to have commented on. The most memorable episode in the novel is Griffin's recollection to Kemp of how, newly invisible in London, he discovered himself not omnipotent (as he fondly expected) but more wretched than the most destitute of street beggars, wholly impotent, a modern version of Lear's poor forked animal:

'But you begin to realise now', said the Invisible Man, 'the full disadvantage of my condition. I had no shelter, no covering. To get clothing was to forego all my advantage, to make of myself a strange and terrible thing. I was fasting; for to eat, to fill myself with unassimilated matter, would be to become grotesquely visible again. (p. 116)

This sticks indelibly in the mind. Like Midas, Griffin's dream of vast power turns to a terrible curse. It is January. What more distressing than to be naked and starving in the cold streets?

And yet, if we think about it, Griffin could have been as comfortably covered and as well fed as any of his visible fellow-Londoners. As he tells Kemp, describing his first experiments, he discovered early on that any fibre, vegetable, or woody matter can be rendered invisible—particularly if it has not been died or stained.[2] Griffin proves this in his earliest experimental trials:

I needed two little dynamos, and these I worked with a cheap gas engine. My first experiment was with a bit of white wool fabric.

It was the strangest thing in the world to see it in the flicker of the flashes soft and white, and then to watch it fade like a wreath of smoke and vanish. (p. 96)

Griffin moves on to his neighbour's white cat, and the white pillow on which the animal's drugged body is lying—both of which are rendered invisible by his little gas-powered machine.

It is clear that, with a little forethought, Griffin could quite easily make himself an invisible white suit of clothing. He could also render food invisible before eating it, so that its undigested mass did not show up in his otherwise transparent entrails. He could, if he were patient enough, construct himself an invisible house out of invisible wood. If due to the invasions of suspicious neighbours, he had no time to do this in London he might certainly do it during his many weeks at Iping (where, as he tells Kemp, most of his efforts seem to be vainly directed towards finding a chemical formula which will enable him to be visible or invisible at will).

The naked, starving, unhoused Griffin would, logically, seem to be that way not because of any fatal flaw in his science. He could, as has been said, walk around in an invisible three-piece suit, with an invisible top hat on his invisible head, an invisible umbrella to keep himself dry, and an invisible three-course meal in his belly. That he does not avail himself of these amenities may, conceivably, be ascribed to a mental derangement provoked by the excruciating pain of the long dematerialization process.

I had not expected the suffering. A night of racking anguish, sickness and fainting. I set my teeth, though my skin was presently afire; all my body afire; but I lay there like grim death. I understood now how it was the cat had howled until I chloroformed it. Lucky it was I lived alone and untended in my room. There were times when I sobbed and groaned and talked.

But I stuck to it. I became insensible and woke languid in the darkness. (p. 101)

But, closely examined, Griffin's derangement seems to originate in the condition of his life well before his agonizing passage into invisibility. His aggrieved sense of alienation evidently began early, probably with the bullying and jeers he attracted as a child, on account of his physical abnormality. It is his physical repulsiveness that strikes those who remember him as an adult. Kemp recalls Griffin at University College as 'a younger student, almost an albino, six feet high, and broad, with a pink and white face and red eyes—who won the medal for chemistry'. This evidently was six or seven years since. In his last years as a student, and in his first employment as a lecturer in an unfashionable provincial university, Griffin (who, one guesses, has neither men nor women friends) has deteriorated into a condition of clear paranoia:

I kept it [i.e. his discovery] to myself. I had to do my work under frightful disadvantages. Oliver, my professor, was a scientific bounder, a journalist by instinct, a thief of ideas,—he was always prying! And you know the knavish system of the scientific world. I simply would not publish, and let him share my credit. I went on working. I got nearer and nearer making my formula into an experiment, a reality. I told no living soul, because I meant to flash my work upon the world with crushing effect,—to become famous at a blow . . . To do such a thing would be to transcend magic. And I beheld, unclouded by doubt, a magnificent vision of all that invisibility might mean to a man,—the mystery, the power, the freedom. Drawbacks I saw none. You have only to think! And I, a shabby, poverty-struck, hemmed-in demonstrator, teaching fools in a provincial college, might suddenly become—this. (pp. 93–4)

Griffin, in his maniac delusions of divine superiority, despises humanity. When, therefore, he goes among the London crowds naked and starving it is because, like the self-divested Lear on the heath, he is deliberately refusing to wear the uniform of his herd-like fellow man. He has chosen to strip himself. Nakedness is the sign of his difference, and his godlike superiority over the lesser, visible beings, he despises. He no more needs trousers than Jove or Satan. It is beneath his notice to concern himself with such minutiae.

The Oxford World's Classics *The Invisible Man* is edited by David Lake with an introduction by John Sutherland. (For copyright reasons, this volume is only available in the United States.)

Bram Stoker · *Dracula*

Why does the Count come to England?

Our obsession with vampires supports a commercial empire which mass-produces books, films, cartoons, television dramas, comics, and novelties (such as plastic vampire teeth and Bela Lugosi capes). Like the great Transylvanian monster himself, it would seem that the vampire industry cannot die. Just when you think it is finally exhausted, along come a couple of big-budget, millions-at-the-box-office-earning movies like Francis Ford Coppola's *Bram Stoker's Dracula* (1993) and the Anne Rice-originated, Neil Jordan-directed, Tom Cruise-starring *Interview with the Vampire* (1995).

The essential book on the vampire *cultus* is Paul Barber's *Vampires, Burials and Death.*[1] Barber's survey is massively debunking and wholly convincing, drawing as it does on the resources of folklore scholarship, anthropology, mythography, and forensic pathology (particularly the evidence of post-mortem medical investigators and autopsies). Barber shows how superstitions about vampires—which are found in cultures as remote from Transylvania as China—originate not in the epic misdeeds of Vlad the Impaler, but in the behaviour of the human corpse after death. The dead body is not inert, but a veritable hothouse of chemical and physiological activities. It moves, makes noises, and excretes fluids. This post-mortem activity, Barber suggests, is rationalized by primitive peoples into the vampire (or 'undead') myth.

Dracula, as A. N. Wilson pointed out in his World's Classics introduction, is a mishmash of elements picked up from the author's experience in the popular theatre (exactly at that moment when it was about to transmute into the cinema industry) and from Gothic predecessors such as John Polidori's *The Vampyre* (1819), J. M. Rymer's *Varney the Vampire* (1847), and Sheridan Le Fanu's superior novella, *Carmilla* (1872). On to this Stoker pasted some new-fangled psychiatric theory, derived from the French alienist Charcot, one of Freud's main precursors (see p. 191, where Van Helsing indicates he is a disciple). Overlaying the whole work is the kind of paranoid anxiety induced by the 'invasion fantasies' which (following Colonel George Chesney's *The Battle of Dorking* in 1871) were a popular fictional genre at the end of the century.[2] Like Wells's Martians (which also sent shivers down English spines in 1897), Stoker's monstrous vampire is a deadly and alien invader, bent on destroying England's green and pleasant land (both the Martians and Dracula support themselves on a diet of English blood, interestingly enough).

As its title proclaims, Francis Ford Coppola's sumptuously produced *Bram Stoker's Dracula* prides itself on being more authentic than its predecessors. Coppola strips away the superstructural mythologies which originate in F. W. Murnau's 1922 film, *Nosferatu*; principally the convention (which is not found in Stoker) that Dracula is destroyed by sunlight. Coppola observes Stoker's less florid conception, which is that Dracula can only exercise his superhuman powers fully at night. During the day he is obliged to take a mundane, human form, and his wings are correspondingly clipped. But he is quite capable of moving around by daylight with the freedom of any other gentleman. Traditionally, film-makers have loved the Murnau final twist of Dracula surprised by sunlight and turning first to gorgonzola and then to bones and sawdust. It makes wonderful cinema. When the first (of six) Hammer versions of *Dracula* came out in 1958, the last sequence, in which Christopher Lee deliquesced in a shaft of morning sunshine, was thought so horrible that the British censor demanded it be cut.

None the less, Coppola's *Dracula* deviates from Stoker's original text in two important ways. First, by inventing a wildly romantic reason for Dracula's coming to England (he sees in Mina the reincarnation of the wife whom he loved and lost while still human: it is this undying love that brings him to England). Secondly, Coppola throws the film back into a 'Gothic' nineteenth-century England, lush as a Leighton oil-painting, but essentially as ahistorical as Ruritania.

The key to reading *Dracula* and recovering Stoker's artistic intention is, I would suggest, close attention to the large number of spikily contemporary references in the text to recent gadgetry, communications technology, and scientific innovation. It is significant, for example, that Jonathan Harker records his journal in shorthand (p. 1). Later, he refers in passing to his 'Kodak', with which he has photographed the English estate in which Dracula is interested (p. 23). Mina, we are told, is learning to 'stenograph', so that when she marries Jonathan she can be his 'typewriter girl' (p. 53). There are numerous references to the New Woman vogue, something that peaked in 1894 (Mina, although an advanced member of her sex, draws the line at aligning herself with New Women, what with their outrageous 'open sexual unions'; see, for example, p. 89). Lucy Westenra's life is prolonged, but not saved, by a blood transfusion (this is one of many references to up-to-the-minute medical advances—see, for example, the reference to brain surgery, p. 276). Lucy's phonograph cylinders are used by Dr Seward to make memoranda (p. 142). Van Helsing even develops an early version of radar employing Mina's powers as a mesmeric medium to locate the fleeing monster.

On his part, Dracula hates modernity—or, at least, he is nervous of it. He cannot read shorthand and throws Harker's encrypted writings on the fire in disgust. He chooses to come to England by sail, not steamboat. He studiously avoids the railway for the transport of his earth-filled boxes, choosing instead gipsy carts. What this means is that in the struggle between Van Helsing and Dracula, we have a contest between the 'pagan world of old' and 'modernity' (p. 134). A demon from the Dark Ages pitted against men of the 1890s armed with Winchester rifles, telegrams, phonographs, modern medicine and science. Stoker's Transylvania is certainly Gothic and ahistorical. But his England is as up-to-date as that week's edition of *Tit-bits.*

Why, it may be asked, does Dracula want to come to England? It would seem he has something more than tourism in mind. When we first encounter him, through Harker; he is practising his English to make it flawless, and is studying

> books . . . of the most varied kind—history, geography, politics, political economy, botany, geology, law—all relating to England and English life and customs and manners. There were even such books of reference as the London Directory, the 'Red' and 'Blue' books, Whitaker's Almanack, the Army and Navy Lists, and—it somehow gladdened my heart to see it—the Law List. (p. 19)

One apprehends from this that Dracula does not want to visit England—he wants to invade it, conquer it, make it his own infernal kingdom. It is notable that his activities in England are very different from those in Transylvania where, apparently, his depredations on the local populace are random, infrequent and rather circumspect (the only local victim we learn of is one baby). In Transylvania Dracula is apparently careful about propagating his kind, keeping his retinue of undead companions to a handful. But in England his promiscuity triggers off a potential infectious epidemic. Lucy becomes one of the undead, and as the 'Bloofer Lady' promptly embarks on infecting any number of children who in their turn will infect others. In a year or so, we can calculate, England will be a pest-hole.

There are huge risks in Dracula moving from his castle fastness in Transylvania. The business of the fifty boxes makes him very vulnerable. The journey itself involves what would seem to be unacceptable risks—his ship is almost wrecked off Whitby (and death by drowning is, together with a stake through the heart, one of the sure ways in which Dracula can be exterminated). Why do it, and if he must do it, why not choose Germany, which at least would shorten the distance back to his lair and would not entail passing over the dangerous element of water?

The reason for Dracula's coming to England is divulged late in the narrative by Van Helsing. 'Do you not see', he asks Harker, 'how, of late, this monster has been creeping into knowledge experimentally?' (p. 302). Dracula, in other words, is learning how to think scientifically. The point is elaborated by the perspicacious professor a little later:

> With the child-brain that was to him he have long since conceive the idea of coming to a great city. What does he do? He find out the place of all the world most of promise for him. Then he deliberately set himself down to prepare for the task. He find in patience just how is his strength, and what are his powers. He study new tongues. He learn new social life; new environment of old ways, the politic, the law, the finance, the science, the habit of a new land and a new people who have come to be since he was. His glimpse that he have had whet his appetite only and enkeen his desire. Nay, it help him to grow as to his brain; for it all prove to him how right he was at the first in his surmises. He have done all this alone; all alone! from a ruin tomb in a forgotten land. What more may he not do when the greater world of thought is open to him? (pp. 320–1)

Dracula, we apprehend, has chosen England because it

is the most modern country in the world—the most modern that is, in its social organization, its industry, its education, its science. Put in its most banal form, he has come to England to learn how to use the Kodak, how to write shorthand, and how to operate the recording phonograph in order that he may make himself a thoroughly modern vampire for the imminent twentieth century.

The Oxford World's Classics *Dracula* is edited by Maud Ellmann.

Rudyard Kipling · *Kim*

How old is Kim?

Alan Sandison notes in his introduction to the World's Classics edition of *Kim* that 'As a physical being, Kim remains a rather shadowy figure' (p. xvii). One of the shadowier aspects of Kipling's young hero is his precise age during the first half of the narrative—the period of his Indian liberty before he is (reluctantly) made a 'Sahib'. The opening encounter finds Kim—indistinguishable from Hindu lads of his age—playing 'King of the Castle' astride the great gun Zam-Zammah (a symbol of the mastery of the Punjab, as Sandison points out). 'King of the Castle' is, of course, child's play. The brief description which Kipling gives of the game suggests the players are very childish indeed:

> 'Off! Off! Let me up!' cried Abdullah, climbing up Zam-Zammah's wheel.
>
> 'Thy father was a pastry-cook, Thy mother stole the *ghi*,' sang Kim. 'All Mussalmans fell off Zam-Zammah long ago!'
>
> 'Let *me* up!' shrilled little Chota Lal in his gilt-embroidered cap. His father was worth perhaps half a million sterling, but India is the only democratic land in the world.
>
> 'The Hindus fell off Zam-Zammah too. The Mussalmans pushed them off. Thy father was a pastry cook—' (p. 4)

At this point, Kim has his first astonishing sight of the Lama, whose *chela*, or guide, he is to become.

In passing we are given the bare details of Kim's parentage. His father was a young colour-sergeant in the Mavericks, an Irish regiment. Kimball O'Hara married a nursemaid in a Colonel's family (given the shortage of eligible young European women in India, Sergeant O'Hara must have been a dashing fellow). After marriage, he resigned the service and took a post on the Sind, Punjab, and Delhi railway. A child arrived soon after and soon after that Annie O'Hara died of cholera in Ferozepore. The husband, left with a 'keen three-year-old baby', went to the bad, took to opium, and died. From his very imperfect English (he thinks, for example, that his father served in an 'Eyerishti' regiment, p. 86), Kim must have been left a young orphan. He has been brought up by a lady of easy virtue in the bazaar.

The keynote in early descriptions of Kim, 'the little friend of the world', is the incantatory use of the term 'little'. Take, for example, the early instance of his street-wise resourcefulness in kicking away the bull (a sacred beast to Hindus) from the stall of the vegetable seller, thus ensuring a charitable donation of food for the Lama:

> The huge, mouse-coloured Brahminee bull of the ward was shouldering his way through the many-coloured crowd, a stolen plantain hanging out of his mouth. He headed straight for the shop, well knowing his privileges as a sacred beast, lowered his head, and puffed heavily along the line of baskets ere making his choice. Up flew Kim's hard little heel and caught him on his moist blue nose. He snorted indignantly, and walked away across the tram rails, his hump quivering with rage. (p. 14)

'Hard little heel' suggests a hard little fellow. When he falls in with the Mavericks, the chaplain (seizing on the charm around his neck which contains his birth certificate) tells Kim: 'Little boys who steal are beaten. You know that?' (p. 84). A couple of pages on, in the same vein, the narrative observes that 'small boys who prowl about camps are generally turned out after a whipping. But [Kim] received no stripes' (p. 86). On the next page he is 'a phenomenal little liar' (p. 87) and a 'little imp' (p. 96) and a 'little limb of Satan' (p. 97).

Answering from their general impressions, most readers would see Kim as somewhere between 9 and 12. There is some confirmation of this when we are told that 'Kim had many dealings with Mahbub *in his little life*,—especially between his tenth and his thirteenth year' (p. 18; that is, between the ages of 9 and 12). When Kim falls into the hands of his father's regiment, it is clear that he is too 'little' to be recruited as a drummer-boy—young soldiers who are 14 and up. One (later discarded) plan is to send the little waif 'to the Military Orphanage at Sanawar where the regiment would keep you till you were old enough to enlist' (pp. 104–5). Kim could, of course, enlist on his fourteenth birthday, as have other drummer-boys in the regiment. As the son of a not very illustrious colour-sergeant, this would seem quite appropriate.

It is fair to say that most readers see the Kim of Kipling's early chapters as a 12-year-old urchin, rather young and small for his years. It is partly because he is so 'little' that the Mavericks adopt him as a mascot, to be treated with unusual care. Hollywood, which has to be definite about such things, made Kim the same very juvenile age as Sabu the Elephant boy, or 'Boy' in the Tarzan films, in the 1950 film (starring Errol Flynn as Mahbub, and the child-star Dean Stockwell as Kim).

Perplexingly, a different calculus of Kim's age emerges in the Maverick scenes—one which would make him significantly older. On the face of it, one can see why this happens. It is specifically stated in the first chapter that Sergeant Kimball O'Hara was demobilized from his regiment, which duly 'went home without him' (p. 1). Kim cannot have been born long after—at most a couple of years and conceivably less. What this means is that the Mavericks have been posted home from India to Ireland, and then posted out again. Moving a thousand men and all their *matériel* twice across the globe is no small thing, even for the rulers of the British Empire, and two such tours of duty in peacetime could not possibly have taken place without an interval of many years, and possibly decades. Father Victor; who was evidently a young Catholic chaplain on the earlier Indian posting, recalls 'I saw Kimball married myself to Annie Shott' (p. 86). The impression the reader gets is that it happened a long time ago.

By this second chronological scheme, Kim is something over 14—on the brink of adolescence. It is, of course, hard to square this with his playing 'King of the Castle' and flashing his 'hard little heel' at the Brahminical bull. But, by reference to specific date-markers in the book, we can calculate that Kim spends something under three years at St Xavier's, where he makes remarkable progress:

> It is written in the books of St Xavier in Partibus that a report of Kim's progress was forwarded at the end of each term to Colonel Creighton and to Father Victor, from whose hands duly came the money for his schooling. It is further recorded in the same books that he showed a great aptitude for mathematical studies as well as map-making, and carried away a prize (*The Life of Lord Lawrence*, tree-calf, two vols., nine rupees, eight annas) for proficiency therein; and the same term played in St Xavier's eleven against the Allyghur Mohammedan College, his age being fourteen years and ten months . . . Kim seems to have passed an examination in elementary surveying 'with great credit', his age being fifteen years and eight months. From this date the record is silent. His name does not appear in the year's batch of those who entered for the subordinate Survey of India, but against it stands the words, 'removed on appointment.'
>
> Several times in those three years, cast up at the Temple of the Tirthankers in Benares the lama, a little thinner and a shade yellower, if that were possible, but gentle and untainted as ever. (pp. 164–5)

It is earlier indicated that graduation from the school will happen when Kim is 17 (pp. 118, 176), which indicates that he enters at 14, or something over.

The cricket-playing Kim seems significantly older than the King-of the-Castle-playing Kim of eight months earlier ('little Kim', as we may call him). Possibly, as young boys do, Kim had his growing spurt at this period, and started on the dramatic physical changes and enlargements involved in puberty.[1] What seems more likely is that Kipling has two Kims in his mind. One, 'little Kim', corresponds to the 'Indian' Rudyard, who left India at the age of 6, and who idealized his early experience of the subcontinent in Kim's early escapades in the bazaar. Kipling saw this early segment of his life as immensely important. He begins his autobiography, *Something of Myself*, with the statement: 'Give me the first six years of a child's life, and you can have the rest.'[2] That formative phase of his life would seem to be memorialized in the early chapters of *Kim*, and Kim is correspondingly infantile.

The other Kim corresponds to the pubescent Kipling who was enrolled at the age of 13 in the United Services College at Westward Ho!, in Devon (an institution with clear similarities to St Xavier's—both institutions train English boys for the colonial service). One Kim is a diminutive urchin, the other a coltish schoolboy. A third, and more stable Kim, the adolescent on the verge of manhood, dominates the second half of the novel. How old is Kim? It depends on the angle.

The Oxford World's Classics *Kim* is edited by Alan Sandison.

Notes

Mansfield Park

1. Edward Said, *Culture and Imperialism* (London, 1993).
2. Kotzebue's play, 'The Natural Son' was first produced in England, translated as *Lovers' Vows*, 1798–1800. Mrs Inchbald's translation (which is presumably what is used by the amateur troupe at Mansfield Park) is printed as a supplement to R. W. Chapman's edition of the novel, in *The Novels of Jane Austen*, 5 vols. (Oxford, 1934). *Mansfield Park* is volume 3 in the set.
3. According to Chapman, the composition of *Mansfield Park* was begun 'about February 1811' and finished 'soon after June 1813'.
4. The World's Classics edition of Thomas Hughes's *Tom Brown's Schooldays* is edited by Andrew Sanders.
5. In an appendix on the chronology of the novel, Chapman deduces (from internal evidence) that Austen used almanacs of 1808–9 in order to arrange day-to-day, month-to-month events and episodes in the novel. But she did not necessarily identify the *historical* period as that year. Chapman sees the question of historical dating as ultimately unfixable: 'As to the "dramatic" date of the story, the indications are slight. It is probably hopeless to seek to identify the "strange business" in America. Many strange things happened in those years. The *Quarterly Review* was first published in 1809 (and therefore could not have been read at Sotherton in 1808). Crabbe's *Tales* (1812) are mentioned. A state of war is implied throughout, and there is no mention of foreign travel, except Sir Thomas's perilous voyage' (p. 556).
6. See chapter 13 of Lockhart's *Memoirs of the Life of Sir Walter Scott, Bart.* (London, 1836, much reprinted) describing Scott's removal to the small farm of Ashestiel, and the advantageous economic arrangement it represented.
7. *Mansfield Park*, Stephen Fender and J. A. Sutherland, Audio Learning Tapes (London, 1974).
8. For the role of evangelicalism in *Mansfield Park* see Marilyn Butler, *Jane Austen and the War of Ideas* (Oxford, 1975), 219–49; and the same writer's essay 'History, Politics, and Religion' in

(ed.) J. David Grey, *The Jane Austen Handbook* (London, 1986). Professor Butler convincingly argues that the evangelicals of the early nineteenth century are not to be confused with the lower-class evangelicals of the Victorian period; 'During the war against France and against "revolutionary principles", pressure for a renewed commitment to religious and moral principle was not so much petit bourgeois as characteristic of the gentry' (p. 204).

9. The whole business of 'Slavery and the Chronology of *Mansfield Park*' was revived in a lively article by Brian Southam in the *TLS* ('The Silence of the Bertrams'), 17 Feb. 1995. The key assumption in Southam's argument is that the 1812 publication of Crabbe's *Tales* and the information that Sir Bertram returns in October enables us to 'pinpoint the course of events'. The chronology Southam deduces is as follows: 'Sir Thomas and Tom leave for Antigua about October 1810; Tom returns about September 1811; Sir Thomas writes home, April 1812; Fanny in possession of Crabbe's *Tales*, published September 1812; Sir Thomas returns, late October 1812; Edmund turns to Fanny, summer 1813' (p. 13). Southam correlates this schedule of events with the evolution of the slave trade, following the Abolition Act of 1807. Convincing as Southam's argument is, I remain somewhat sceptical that Jane Austen would expect her readers to register September 1812 as the publication date of Crabbe's *Tales* (much of his poetry has an earlier publication date), and the detail may have been as loosely installed in her mind as in those of the bulk of her readers, who would simply recall the volume as a 'recent' publication. It may also be, as I argue above, that '*Tales*' should be read as the less precisely dateable 'tales'.

Waverley

1. *The Letters of Sir Walter Scott*, ed. Sir Herbert Grierson (London, 1932–7), iii. 478–9.
2. John Hill Burton, *The History of Scotland from the Revolution to the Extinction of the Last Jacobite Insurrection, 1698–1748*, 2 vols. (London, 1853), ii. 463.
3. See chapter 7 of Lockhart's *Life of Scott*.

Emma

1. These essays of Scott's are conveniently collected in Ioan Williams (ed.), *Sir Walter Scott on Novelists and Fiction* (London, 1968).
2. See Edgar Johnson, *Sir Walter Scott: The Great Unknown*, 2 vols. (London, 1970), ii. 1084.
3. See the World's Classics edition of *King Solomon's Mines*, edited by Dennis Butts, p. 332.
4. See the World's Classics edition of *The Swiss Family Robinson*, edited by John Seelye, pp. 25, 332.
5. R. W. Chapman (ed.), *Emma* (London, 1933), 493.
6. Presenting to the mind's eye a montage of the year's passing season was a favourite device of William Cowper, a poet Austen is known to have read. See for instance vi. 140–60 of *The Task*, 'But let the months go round, a few short months . . .'

Frankenstein

1. One text reprinted in World's Classics (eds. James Kinsley and M. K. Joseph) is the revised, 1831 'third edition'. The other from which I have taken quotations, is 'the 1818 text', edited by Marilyn Butler. Substantive changes between the 1818 and revised 1831 texts are noted in Appendix B of Professor Butler's edition.
2. T. J. Hogg, *The Life of Percy Bysshe Shelley* (London, 1858), i. 70–1.
3. Maurice Hindle (ed.), *Frankenstein* (Harmondsworth, 1988), p. xxi.
4. See Steven E. Forry, *Hideous Progeny: Dramatizations of Frankenstein, from Mary Shelley to the Present* (Philadelphia, 1990), p. ix. Forry detects three phases in the successive dramatizations of the story: (1) 1823–32 (which saw fifteen versions) were years of 'transformation and proliferation' during which 'the myth was mutated for popular consumption'; (2) 1832–1900 were 'years of diffusion', in which the myth was spread into the general Anglo-American consciousness; (3) 1900–30 were 'years of transition', as the stage-generated versions of *Frankenstein* were gradually displaced by imagery derived from the cinema. For the changing cultural fortunes of Victor Frankenstein and his monster (frequently the two were confused) in the hundred years following 1818, see Chris Baldick, *In Frankenstein's Shadow* (Oxford, 1987).

5. Edison's *Frankenstein* was presumed lost, but a print was recovered in the 1980s. See Forry, *Hideous Progeny*, 80.
6. Ibid. 85.
7. *Mary Shelley's Frankenstein*, Leonore Felisher, based on a screenplay by Steph Lady and Frank Darabont, from the novel by Mary Shelley, with an afterword by Kenneth Branagh (New York, 1994), 307.
8. Lightning had only recently been identified as an electrical phenomenon by Benjamin Franklin. See the experiment described on p. 24 of *Frankenstein* and the note to it on p. 255.
9. The role of Fritz goes back to the most successful of the early stage adaptations, *Presumption, or the Fate of Frankenstein* (1823), by Richard Brinsley Peake. The play is usefully reprinted in the Everyman edition of *Frankenstein*, ed. Paddy Lyons (London, 1994).
10. The significance of the changes which Mary Shelley made between the 1818 and 1831 texts is examined by Marilyn Butler in 'The First *Frankenstein* and Radical Science', *TLS (*9 Apr. 1993), 12–14. Professor Butler explains the relevance of the first edition to the 'celebrated publicly staged debate of 1814–19 between two professors at London's College of Surgeons [John Abernethy and William Lawrence] on the origins and nature of life, now known as the vitalist debate' (p. 12). Mary Shelley toned down her opinions in the 1831 revised text of *Frankenstein*. See also the appendices and introduction to Professor Butler's World's Classics edition of the novel.
11. Ellen Moers, 'Female Gothic', in *New York Review of Books* (21 Mar. 1974), reprinted in G. Levine and U. Knoepflmacher (eds.), *The Endurance of Frankenstein* (Berkeley: California, 1979), 77–87.
12. For a survey of recent feminist argument and discussion on the novel see Catherine Gallagher and Elizabeth Young, 'Feminism and *Frankenstein*: A Short History of American Feminist Criticism', *The Journal of Contemporary Thought*, 1 (Jan. 1991), 97–109. An influential reading along this line is found in Anne K. Mellor, *Mary Shelley: Her Life, her Fiction, her Monsters* (Berkeley: California, 1988). According to Mellor: 'From a feminist viewpoint, *Frankenstein* is a book about what happens when a man tries to have a baby without a woman'. See also Sandra M. Gilbert's and Susan Gubar's *The Madwoman in the Attic* (New

York, 1979) which reads *Frankenstein* as an Eve-myth. A good selection and discussion of feminist readings of the novel is given in the endmatter of Lyons's Everyman edition.

Oliver Twist

1. Jonathan Grossman, in a forthcoming article in *Dickens Studies Annual*, points out that this is the first time Oliver refers to Fagin as 'The Jew'—a fact of which he seems previously to have been unaware.
2. Bayley's essay, 'Things as they are', is to be found in John Gross and Gabriel Pearson (eds.), *Dickens and the Twentieth Century* (London, 1962), 49–64.
3. J. Hillis Miller, *Dickens: The World of his Novels* (Cambridge: Mass., 1958).
4. Colin Williamson, 'Two Missing Links in *Oliver Twist*', Nineteenth-Century Fiction, 22 (Dec. 1967), 225–34.
5. See Alison Winter; 'Mesmerism and Popular Culture in early Victorian England', *History of Science*, 32 (1994), 317–43. Kaplan discusses the O'Key connection at length in *Dickens and Mesmerism*, 34–44.
6. Kathleen Tillotson (ed.), *Oliver Twist* (Oxford, 1966), 392. Thackeray, who illustrated his own serial novels, frequently ran into difficulties supplying designs sufficiently ahead of time for the engraver. See the 'Commentary on Illustrations' by Nicholas Pickwoad, in *Vanity Fair*, ed. P. L. Shillingsburg (London and New York, 1989), 643–4. Shillingsburg gives further examples in his companion edition of *Pendennis* (London and New York, 1990), 'Writing and Publishing *Pendennis*', 375–98.
7. That Dickens altered his plan of *Oliver Twist* as he went along is convincingly argued by Burton M. Wheeler, 'The Text and Plan of *Oliver Twist*', Dickens Studies Annual, 12 (1983), 41–61. Wheeler notes particularly that 'the conspiracy between Monks and Fagin does not bear close scrutiny' (p. 56) and that this section of the novel bears witness of being improvised from one instalment to the next.
8. See *The Letters of Charles Dickens, vol. 1, 1820–39*, ed. Madeline House and Graham Storey (Oxford, 1965), 403, 461.

Martin Chuzzlewit

1. George Eliot's schedule for these events is to be found, with some alterations and later changes of mind, in her 'Quarry for *Middlemarch*', A. T. Kitchel (ed.) (Los Angeles, 1950), 45–6. As Kitchel's transcription shows it was, for instance, Eliot's original intention to have Dorothea marry as early as 1827.
2. I examine this topic in 'The Handling of Time in *Vanity Fair*', Anglia, 89 (1971), 349–56.
3. See P. D. Edwards, 'Trollope's Chronology in *The Way We Live Now*', Notes and Queries, 214 (1969), 214–16.
4. See Bert Hornback, 'Anthony Trollope and the Calendar of 1872: The Chronology of *The Way We Live Now*', Notes and Queries, 208 (1963), 454–8. Peter Edwards's article in *N&Q*, cited above, specifically refutes Hornback's thesis.
5. Dickens is also vague about historical dating. In *Martin Chuzzlewit* (London, 1985, Unwin Critical Library), 35, Sylvère Monod notes the contradiction of the world of coaching around Salisbury and the steam-driven vessels on which Martin and Mark travel to and in America.
6. Monod makes the point that errors and awkwardness in the narrative construction of *Martin Chuzzlewit* are at least partly to be attributed to the novelist's artistic inexperience (see p. 141). Monod cites as examples of loose threads in the story such things as Pecksniff's loss of all his fortune before Tigg has any reasonable chance of acquiring it (p. 81).

Wuthering Heights

1. Heathcliff's complicated legal manœuvre is explained in Appendix VI, 'Land Law and Inheritance', by E. F. J. Tucker, in *Wuthering Heights*, ed. Hilda Marsden and Ian Jack (Oxford, 1976), 497–9. Tucker assumes there may have been same 'dishonesty' and collusion between Heathcliff and Mr Green, the Gimmerton attorney, in alienating Hindley so entirely from his inhentance. As Tucker points out, Green evidently 'misinforms' Edgar Linton as to the state of the law, so as to favour Heathcliff's claims. Bribery is implied.

Jane Eyre

1. See Michael Mason's note to this episode in his Penguin Classics edition of *Jane Eyre* (1995). Mason quotes George Troup, reviewing the novel in *Tait's Edinburgh Review* (May 1848): 'the voice has not got a telegraphic communication direct to the ear at fifty miles distance, although intelligence by the magnetic wire may travel hundreds and thousands "in no time".'
2. On her visit to London in June 1850 Thackeray made a point of introducing Charlotte Brontë to Catherine Crowe, 'the reciter of ghost stories'. It would be interesting to know how the two ladies got on. See Clement Shorter (ed.), *The Brontës' Life and Letters*, 2 vols. (London, 1908), ii. 94, 147.
3. Winter is drawing on a report by John Elliotson in his journal of mesmeric science, *Zoist*, 2 (1844–5), 477–529, 'Reports of Various Trials of the Clairvoyance of Alexis Didier Last Summer in London'.
4. The connection was via Charlotte's father, Patrick. See Juliet Barker, *The Brontës* (London, 1994), 401, 511.
5. See the *Athenaeum* (30 Nov., 14 Dec., 21 Dec., 1844), 1093–4; 1117–18; 1173–4.

Vanity Fair

1. Keith Hollingsworth in *The Newgate Novel* (Detroit, 1963), 212–15, argues that *Vanity Fair* is, in its last pages, emulating the Newgate Novel. Hollingsworth accepts that 'Rebecca murders Joseph Sedley' and that she does it by poison, presumably arsenic, the poison of choice for murderesses in the 1840s (p. 212).

Henry Esmond

1. Gordon S. Haight (ed.), *The George Eliot Letters*, 9 vols. (London, 1954–78), ii. 67.
2. See, for instance, S. J. Kunitz and Howard Haycraft, *British Authors before 1800* (New York, 1952), 391–2: 'Coleridge, Charles Lamb and Leigh Hunt all admired *Peter Wilkins*, and Southey wrote that the winged people of the book "are the most beautiful creatures of imagination that were ever devised." Paltock's book (dedicated to Elizabeth, Countess Northumberland, presumed to be the prototype of his heroine) ran through four editions before

1800 and at least sixteen later. A musical pantomime based on the book was produced in 1800 and a play in 1827.'

3. Thackeray began his reading for the 'English Humourists' in December 1850. The first series of lectures were delivered in London, 22 May–3 July 1851. Composition of *Henry Esmond* began in August 1851. He delivered lectures around the UK over the following year. The novel was published in October 1852, on the eve of Thackeray's departure to lecture on the English Humourists in America.
4. See G. N. Ray, *Thackeray: The Uses of Adversity* (New York, 1955), 1–3. Thackeray's actual injunction to his daughters was 'Mind, no biography!'

Bleak House

1. 'Mud-pusher' was a slang term for crossing-sweepers in the Victorian period. See J. Redding Ware, *Passing English of the Victorian Era: A Dictionary of Heterodox English, Slang, and Phrase* (London, 1909), 178.
2. F. S. Schwarzbach, *Dickens and the City* (London, 1979), 124.
3. Norman Gash, *Robert Surtees and Early Victorian Society* (Oxford, 1993), 48–9. On the question of 'little commercial value' that Gash mentions, Susan Shatto, in her *Companion to Bleak House* (London, 1988) cites from Mayhew the fact that 'great quantities of mud were collected daily and carted off to manure barges, the owners of which sold each load for £5 to £6' (p. 23). Mayhew estimated that four-fifths of the gathered-up street dirt was horse droppings and cattle dung, and around 100 tons of the substance was deposited on the London streets every working day.
4. H. Mayhew, *London Labour and the London Poor* (London, 1851, repr. and enlarged, 1861), ii. 465. The long section on crossing-sweepers extends to ii. 507.
5. Harvey P. Sucksmith disputes Gill's assertion in 'The Dust Heaps in *Our Mutual Friend*', Essays in Criticism, 23 (1973), 206–12.
6. See the World's Classics edition of *Middlemarch*, ed. David Carroll, 72, 74.

Villette

1. See, for instance, Terry Eagleton, *Myths of Power: A Marxist Study of the Brontës* (London, 1975), 92; Janet Gezari, *Charlotte*

Brontë and Defensive Conduct: The Author and the Body at Risk (Philadelphia, 1992), 167–70.

2. Kate Millett notes that had Paul returned to claim his bride, 'we should never have heard from her'—Madame Emanuel, in other words, could not have written *Villette*. Brenda Silver astutely notes: 'Lucy's life does not end with Paul's; the observant reader will have noted that the school clearly continues to prosper and that Lucy, by the time she begins her narrative, knows the West End of London as well as her beloved City . . . She has survived the destruction of the romantic fantasy and grown into another reality.' Millett's and Silver's essays are printed in *Critical Essays on Charlotte Brontë*, ed. Barbara Timm Gates (Boston, 1990); see particularly 263, 303.
3. Carol Hanbury Mackay (ed.), *The Two Thackerays: Anne Thackeray Ritchie's Centenary Biographical Introductions to the Works of William Makepeace Thackeray*, 2 vols. (New York, 1988), i. 385.
4. Guinevere Griest, *Mudie's Circulating Library and the Victorian Novel* (Bloomington: Indiana, 1970), i. 21.

Adam Bede

1. *Victorian Studies* (Sept. 1970), 83–9.
2. Sedgwick's essay, which was originally a paper delivered at the MLA annual convention, is printed in *Critical Inquiry*, 17: 4 (Summer 1991), 818–37.
3. Although modern readers frequently complain that Hetty's pregnancy is invisible, Victorians felt that it was obtruded—by precise dating references—too forcibly on the reader. See, for instance, the *Saturday Review* (26 Feb. 1859): 'The author of *Adam Bede* has given in his adhesion to a very curious practice that we consider most objectionable. It is that of dating and discussing the several stages that precede the birth of a child. We seem to be threatened with a literature of pregnancy'. The review is reprinted in *George Eliot: The Critical Heritage*, (ed.), David Carroll, (London, 1971), 73–6.
4. See the chapter 'Effie's phantom pregnancy', above.

The Woman in White

1. Kenneth Robinson, *Wilkie Collins: A Biography* (London, 1951), 149.

2. Nuel P. Davis, *The Life of Wilkie Collins* (Urbana: Illinois, 1956), 216.
3. The review is reprinted in Norman Page (ed.), *Wilkie Collins: The Critical Heritage* (London, 1974), 102–3.
4. Ibid. 124.
5. Ibid. 95.
6. This anomaly in the novel's chronology is noted by W. M. Kendrick, 'The Sensationalism of *The Woman in White*', Nineteenth-Century Fiction, 32: 1 (June 1977), 18–35 (see particularly 23) and by Andrew Gasson '*The Woman in White*: A Chronological Study', *Wilkie Collins Society Journal*, 2 (1982), 12–13.
7. J. G. Millais, *The Life and Letters of John Everett Millais* (London, 1899), 278–9.

Pendennis, A Dark Night's Work, Rachel Ray

1. The text was published in the first serial instalment of the novel, April 1836. The footnote was added for the 'Cheap' 1847 reissue of the novelist's works. See the World's Classics *Pickwick Papers*, 726–7.
2. See *Rob Roy*, chap. 21. For other chronological anomalies see the Everyman edition of *Rob Roy*, ed. J. A. Sutherland (London, 1995).
3. For instance: (1) Pen's visit to Vauxhall in Number 15, where Simpson—who retired in the mid-1830s—still presides; (2) the rage for 'silver fork' novels, which Thackeray pointedly recalls as a foible of 'that time' (p. 525); (3) London excitement at the performance of Taglioni in *The Sylphide* in the early 1830s (p. 478); (4) the huge box-office success of Bulwer's play, *The Lady of Lyons*, in 1838.
4. See J. A. Sutherland, 'Dickens, Reade, and *Hard Cash*', Dickensian, 405, 81, 1 (Spring 1985), 9–10.
5. For a closely examined survey of the composition of *A Dark Night's Work* see J. G. Sharps, *Mrs Gaskell's Observation and Invention* (Arundel, 1970), 353–4.
6. For further minor dating errors in the narrative see ibid. 360. Sharps concludes with something of an understatement, 'Mrs Gaskell's chronology was not especially accurate'.
7. In Ellinor's Italian trip, Mrs Gaskell is recalling her own trip to Rome, Feb.–May 1857. As with her heroine, she was recalled to

England in distressing circumstances. Having made an unhurried return to Paris on 26 May, Mrs Gaskell was met with the news that publication of *The Life of Charlotte Brontë* had been suspended, on grounds of libel. She rushed back to London, in a state of extreme anxiety—all of which is mirrored in the account of Ellinor's return to save Dixon's life. See Winifred Gérin, *Elizabeth Gaskell* (Oxford, 1976), 187–8.

8. The first edition of the World's Classics *Rachel Ray* (1988) had as its cover illustration one of Millais's unused illustrations for the novel, showing Rachel dressed in an extravagant crinoline of 1860s vintage.

Phineas Finn

1. See *TLS* (1944), 156, 192, 372.
2. See Simon Raven, 'The Writing of "The Pallisers"', *The Listener*, 91 (1974), 66–8.
3. Stephen Wall, in *Trollope: Living with Character* (London, 1988), notes that the last-page marriage between Phineas and Mary 'verges on the perfunctory'. It is best read, Wall suggests, as 'an action of Phineas's Irish self' (p. 149).

Middlemarch

1. It is feasible that Farebrother is simply retailing vulgar Middlemarch gossip here (as the narrative earlier suggests), assuming that all pawnbrokers are Jewish. Since Dunkirk attends the same fundamentalist Christian church as Bulstrode he would have to be an apostate as well as Jewish. If the 'grafting of the Jew pawnbroker' is accepted as true it may be taken to have consequences in the text. Bernard Semmel, in *George Eliot and the Politics of National Inheritance* (New York 1994), 97–8, suggests that Ladislaw's revulsion at Bulstrode's offer to make amends may reflect his horror at discovering his unsuspected Jewish ancestry. Semmel, who accepts that Dunkirk was Jewish (and Ladislaw therefore partly Jewish), further suggests that the characterization of Will may owe something to Eliot's conception of Disraeli, who in 1830 would be about the same age as her hero.
2. *Middlemarch*, ed. R. D. Ashton (Harmondsworth, 1994), p. xxi.
3. R. D. Ashton, *G. H. Lewes* (Oxford, 1990), 10–11.

The Way We Live Now

1. See, for instance, Noel Annan's downright statement in his introduction to the Trollope Society edition of *The Way We Live Now* (London, 1992): 'Melmotte is a Jew and the upper class characters say, as they would in those days in real life, plenty of disagreeable things about Jews' (p. xiv). Bryan Cheyette, in *Constructions of 'The Jew' in English Literature and Society, 1875–1945* (Cambridge, 1993), is more discriminating. Cheyette sees Melmotte as emblematic of a 'fixed racial Jewishness' without being himself clearly Jewish. Cheyette notes a number of learned commentators who have misled themselves as to Melmotte's being Jewish (see p. 39).

The Prime Minister

1. See J. A. Sutherland, *The Longman Companion to Victorian Fiction* (London, 1989), 511.
2. Victorian railways were very dangerous by today's standards. The *Annual Register, 1875* (London, 1876), 236, reports that, in 1874, 1,424 passengers were killed and 5,041 injured.

Is he Popenjoy?

1. See *The Annual Register, 1875* (London, 1876), 13.
2. See his letter of 10 March 1875, *The Letters of Anthony Trollope*, ed. N. John Hall (Berkeley, 1983), ii. 653.

The Portrait of a Lady

1. Dorothea Krook, 'Two Problems in *The Portrait of a Lady*', in *The Ordeal of Consciousness in Henry James* (New York, 1962), 357.
2. The Hutton review is usefully reprinted in Roger Gard (ed.), *Henry James: The Critical Heritage* (London, 1968), 93–6.
3. See *The Portrait of a Lady* (World's Classics edition), 628.
4. A selection from Hutton's reviews of the serial and collected *Middlemarch* is given in David Carroll (ed.), *George Eliot: The Critical Heritage* (London, 1971), 286–313.
5. See Robert Tener and Malcolm Woodfield, *A Victorian Spectator: The Uncollected Writings of R. H. Hutton* (Bristol, 1990), 30.
6. See G. S. Haight, *George Eliot: A Biography* (London, 1968). A more hostile depiction of the social pressures on 'Mr and Mrs

Lewes' is given by Marghanita Laski, *George Eliot and her World* (London, 1973), 98: 'the Leweses, if they had not come to believe they were somehow or other married, at least began to forget they were not. In 1873, Lewes felt able to tell a correspondent that he had lived with his mother till he married his "Dorothea".'

Dr Jekyll and Mr Hyde

1. There have been 'at least 69 films' of Stevenson's novella, of which the 1941 Spencer Tracy version is the most famous, and Rouben Mamoulian's 1932 version (starring Fredric March) is regarded as the finest achievement artistically. See V. W. Wexman's essay, 'Horrors of the Body' in William Veeder and Gordon Hirsch (eds.), *Dr Jekyll and Mr Hyde after One Hundred Years* (Chicago, 1988), 283–307. This volume has many illustrations from book versions of the story over the years and stills from dramatic and film adaptations.

The Master of Ballantrae

1. See Mary Lascelles, *The Story-Teller Retrieves the Past* (Oxford, 1980), 70 and the RLS essay on 'The Genesis of *The Master of Ballantrae*', reprinted in the 'Vailima' edition of *The Works of Robert Louis Stevenson* (London, 1922), xiv. 15–19.

The Picture of Dorian Gray

1. p. vii. See Karl Beckson (ed.), *Oscar Wilde: The Critical Heritage* (London, 1970), 67–86, for a representative selection of reviews of *Dorian Gray*. Beckson summarizes the novel's extraordinarily hostile reception on 7–12.

Tess of the d'Urbervilles

1. See *Thomas Hardy: The Critical Heritage*, ed. R. G. Cox (London, 1970), 217–18.
2. Ibid. 212.
3. Tony Tanner, 'Colour and Movement in *Tess of the d'Urbervilles*', Critical Quarterly, 10 (Autumn 1968); repr. in R. P. Draper (ed.), *Hardy: The Tragic Novels* (London, 1975), 182–208. The passage quoted comes on 205.
4. *The 1890s*, ed. G. A. Cevasco (New York, 1993).
5. Ian Gregor, *The Great Web* (London, 1974), 182.

6. *Tess of the d'Urbervilles*, ed. James Gibson (London, 1994), 411.
7. Ibid. 417.

The Speckled Band

1. The story was first published in the *Strand*, Feb. 1892. It is recorded as being the author's own favourite among his works.
2. See *The Annotated Sherlock Holmes*, ed. W. S. Baring-Gould, 2 vols. (New York, 1967), i. 263–6. In a sceptical appendix on 'the deadliest snake in India' the editor concludes that there is no such reptile as an 'Indian swamp adder', and no known snake could kill a human victim in ten seconds, as does Roylott's reptilian assassin. The most satisfactory explanation is that the thing in question is half Gila monster and half Indian cobra. It would be impossible for a snake to clamber up a bell-rope as is claimed in the story, 'for snakes do not climb as many think—by twining themselves around an object; they climb by wedging their bodies into any crannies and interstices, taking advantage of every irregularity or protrusion upon which a loop of the body may be hooked' (i. 265).
3. The point is made in *The Annotated Sherlock Holmes*, i. 249.
4. The Roylott misnaming, which is found in the original *Strand* publication of 'The Speckled Band', has unfortunately been corrected in the World's Classics edition. There is a possibility that the 'error' (if that is what it is) arose from the fact that in the original manuscript of the story, 'Helen Stoner is Helen Roylott, and Dr Roylott is her father' (*The Annotated Sherlock Holmes*, i. 246). I would prefer to see it as a subtlety deliberately introduced by the author. A facsimile of the original Feb. 1892 *Strand* text of 'The Speckled Band' may be found in the Wordsworth Classics edition of *The Adventures of Sherlock Holmes* (London, 1992), 213–29. The 'Miss Roylott' comment is on 219.

Jude the Obscure

1. Kate Millett, *Sexual Politics* (New York, 1969), 130.

Weir of Hermiston

1. 'A Hanging', in *The Collected Essays, Journalism and Letters of George Orwell*, ed. Sonia Orwell and Ian Angus, 4 vols. (London, 1968), i. 45.

The Invisible Man

1. Harris Wilson, *Arnold Bennett and H. G. Wells: A Record* (Urbana: Illinois, 1960), 34–5.
2. This point is made by Jack Williamson, in *H. G. Wells, Critic of Progress* (Baltimore, 1973), 85–6.

Dracula

1. Paul Barber, *Vampire, Burials and Death* (New Haven: Conn., 1988).
2. For a survey of invasion fantasies in the 1890s as reflected in fiction, see I. F. Clarke, *Voices Prophesying War, 1763–1984* (London, 1966).

Kim

1. In Kipling's case, puberty did not bring any great change in his stature. As Kingsley Amis notes: 'Physically, Rudyard Kipling was a small man' and never exceeded five feet six inches in height. See Kingsley Amis, *Rudyard Kipling and his World* (London, 1975), 9.
2. Rudyard Kipling, *Something of Myself* (New York, 1937), 3.